Laser Ablation - Applications and Modeling

Edited by Masoud Harooni

Published in London, United Kingdom

Laser Ablation - Applications and Modeling
http://dx.doi.org/10.5772/intechopen.105251
Edited by Masoud Harooni

Contributors
A.Ph. Ilyushchanka, Anastas Savov Nikolov, Calin Constantin Moise, Chunlan Ma, Liu Hong, Marius Enachescu, Mohammed J. Jader, Oana Andreea Lazar, Wasan H. Jawad

First published in London, United Kingdom, 2024 by IntechOpen
IntechOpen is the global imprint of INTECHOPEN LIMITED, registered in England and Wales, registration number: 11086078, 167-169 Great Portland Street, London, W1W 5PF, United Kingdom

British Library Cataloguing-in-Publication Data
A catalogue record for this book is available from the British Library

Additional hard and PDF copies can be obtained from orders@intechopen.com

Laser Ablation - Applications and Modeling
Edited by Masoud Harooni
p. cm.
Print ISBN 978-1-80355-972-8
Online ISBN 978-1-80355-973-5
eBook (PDF) ISBN 978-1-80355-974-2

For EU product safety concerns:
IN TECH d.o.o., Prolaz Marije Krucifikse Kozulić 3, 51000 Rijeka, Croatia,
info@intechopen.com or visit our website at intechopen.com.

Meet the editor

Dr. Masoud Harooni received both a bachelor's and master's in Mechanical Engineering in his home country where he also worked for a few years before moving to the United States in 2010. After moving, he obtained his Ph.D. in Mechanical Engineering with a minor in Laser Materials Processing at the Research Center for Advanced Manufacturing, Southern Methodist University, Dallas, Texas. Following graduation in May 2014, Dr. Harooni joined a tier 1 original equipment manufacturer (OEM) owned by Honda of America to serve as a laser welding engineer. He also worked as part-time faculty for Indiana University-Purdue University at Indianapolis (IUPUI) where he developed and taught a Modern Manufacturing Processes course. He then began his career with the TRUMPF Smart Factory, Chicago, Illinois, in 2017. He served as the head of the welding product group at TRUMPF for 6 years. He is currently Head of Sales in the West at IPG Photonics. Dr. Harooni is also studying for his MBA.

Contents

Preface

The special characteristics of lasers make them very useful in a variety of industries in different applications. One such application is laser ablation. Different industries such as aerospace, defense, automotive, and medical industries are learning the benefits of laser ablation compared to traditional methods. Therefore, it is important to review and learn about new research in the field.

This book, *Laser Ablation – Applications and Modeling*, includes four chapters that review the application and modeling of laser ablation. Written by experts in the field, this book is organized into two sections on "Applications" and "Modeling".

This book is a useful resource for scholars and specialists in the field.

Masoud Harooni, Ph.D.
Head of Sales, US West,
IPG Photonics,
USA

Section 1

Applications

Chapter 1

Laser-Matter Interaction in the Bulk of Semiconductor and Dielectric

Hong Liu and Chunlan Ma

Abstract

The research in the field of laser-induced materials processing is evolving continuously with new inventions in laser technology. This chapter mainly discusses the relevant physical mechanisms of laser ablation based on laser-matter interaction. Femtosecond laser excitation provides suitable conditions for studying the basic processes in irradiated materials, as compared to the duration of these processes, femtosecond laser pulses are sufficiently short. In the process of laser action on the matter, the thermal mechanism, charge carrier removal, thermal and structural effects, and other processes are extremely complex. The ultrashort laser pulse instantly puts the material in a strong nonequilibrium state characterized by hot electrons and cold ions. After the pulse ends, the electron transfers its energy to the ion through electron phonon coupling in sub-picoseconds. This heats up the phonon bath before the slow thermal effect can reconstruct the material. The electron effect plays an important and possibly dominant role in the laser ablation of nonmetallic solid surfaces. This review first describes the mechanism of laser-matter interaction from the perspective of energy, summarizes the electronic excitation and energy relaxation paths of light on semiconductors and dielectric materials, focuses on the electronic excitation and relaxation mechanisms in laser-induced ionization, desorption, and ablation, and finally analyzes the above-mentioned related processes from the perspective of material structure relaxation.

Keywords: laser-matter interaction, ionization, desorption, ablation, electronic excitation, energy relaxation

1. Introduction

Laser ablation (i.e., laser machining (LM)) as a noncontact, wear-free machining process is a promising method to machine difficult-to-cut materials, such as silicon nitride, silicon carbide, aluminum oxide, and boron nitride.

Surface ablation based on high power ultrashort pulse laser plays an increasingly important role in material processing and device preparation technology [1–3]. Classically, the interaction between laser and matter is described by the Beer Lambert law, which states that the transmittance of monochromatic light passing through a material sample decreases exponentially with the length of the optical path, regardless of the incident intensity. This corresponds to a linear light absorption process, as is commonly assumed for nanosecond laser pulses [4]. Electronic defects are generated by initial laser irradiation, changes in absorption characteristics, and laser energy deposition of photons in the middle and later stages of a given or subsequent laser pulse. The ablation plume generated by laser action effectively shields the sample from the influence of laser [5–8], and the plume can keep transparent to the laser at low flux because any free electrons will recombine with ions before being accelerated to ionization energy, which is easy to be absorbed by free electrons in the plume so that they can quickly heat and then ionize the current neutral substance. When the ionization rate exceeds the recombination rate of free electrons and ions, the number of ionized neutral substances doubles. Laser plasma interactions are no longer characterized by classical light absorption, as various nonlinear effects, such as direct collision and field-induced ionization and dissociation reactions, as well as reverse bremsstrahlung [4, 8–10] have emerged. The plume can be transparent to the laser at low flux because any free electron will recombine with the ion before being accelerated to ionization energy. Scholars have established dual temperature models and multi-plasma coupling models for the physical mechanisms of the interaction between laser and matter mentioned above. But there are still many controversies. The above processes can lead to photothermal effects (injection, ionization, vibration heating, melt discharge, evaporation, and plasma formation) and photochemical (bond breaking and generation of new components) processes, which are used for various laser processing and coherent structure preparation [11].

Based on the electronic process of nanosecond and shorter pulse laser ablation in pure semiconductors and dielectric materials, this chapter constructs the dynamics and dynamics images of laser-induced desorption and ablation of nonmetallic crystal surface, expounds the physical mechanism of laser solid interaction, and then gives the physical process that causes the change of bonding state, particle emission, and final ablation. After providing the ionization, desorption, and ablation mechanisms of laser on dielectric and semiconductor surfaces, we construct ionization, desorption, and ablation models under the irradiation limit of strong and weak laser pulses on solid materials, and qualitatively analyze the dynamic mechanism of material structure relaxation under the action of pulsed laser. Finally, a brief summary of the content of this chapter was provided.

2. The physical mechanism of the laser-matter interaction on the different pulsed laser intensity (low and high)

2.1 Basic principle of laser-matter interaction

Bill Lambert's law gives a macroscopic description of the interaction between laser and material and points out that the transmittance of monochromatic light passing through the material sample decreases exponentially with the length of the optical path, and is independent of the incident intensity, which is a linear optical absorption process [4]. However, in the high flux region at a sufficiently long wavelength, this approximation becomes ineffective, where the ablation plume formed effectively protects the sample from the laser [5–7].

The absorption mechanism of laser varies greatly among different materials. For metal nanoparticles, optical absorption is attributed to the collective oscillation of conduction band electrons in the laser field [12–15]. Due to the transient net charge on the particle surface during the oscillation process, a surface plasma is formed. The surface plasmon absorption band depends on the size and shape of the nanostructure (see **Figure 1**) [12, 14–16]. Irregular nanostructures, such as nanocubes and nanoprisms, can generate wider peaks over a wide wavelength range [13, 15, 16]. For heavily doped semiconductors, optical absorption is based on a similar principle of conduction band electron oscillation, while additional absorption is formed due to the excitation of optical phonons, interband transitions, and the dielectric background of polarization generated by all high-frequency excited tails in the material [17]. For pure semiconductors and dielectric materials, the interaction process between pulsed laser and materials is as follows. When irradiated by low-intensity and long-wavelength lasers, the material has little absorption of laser radiation, which will not be discussed here. If the intensity of the incident laser reaches the ablation threshold or the ultra-short wavelength laser acts on the material so that the material can absorb enough energy, the laser will damage the material. Research shows that the production of high-density plasma is the main reason for the damage of high-purity semiconductor and electrolyte materials. The ultrashort pulse laser causes the plasma structure of extremely high density ($10^{18} - 10^{20} cm^{-3}$) free electrons in the dielectric material, simultaneously accompanied by a large number of ionization phenomena.

2.2 Influence of pulse laser parameters on laser-matter interaction

Laser wavelength(λ), the average pulse laser intensity (I_L) [18] and pulse duration (t_{pulse}) are several main parameters that affect the interaction mechanism between laser and material. For lasers with different pulse durations, the mechanisms leading to energy absorption and target ablation are completely different (**Figure 2**). Research shows that if the duration of the laser pulse on the metal workpiece is shorter, the quality of laser ablation will be improved [20, 21].

Thermal diffusion and direct interaction between laser and material can ablate the material. Due to the fact that thermal energy can diffuse in any direction of the material, related erosion can also occur in any direction. Laser ablation is mainly based

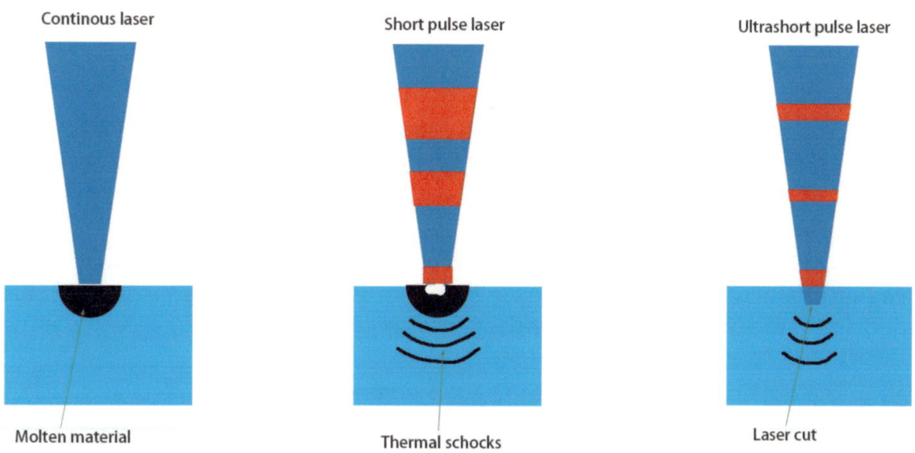

Figure 1.
Laser-material interaction in continuous, short pulse (nanosecond) and ultrashort pulse laser. Reproduced from ref. [19].

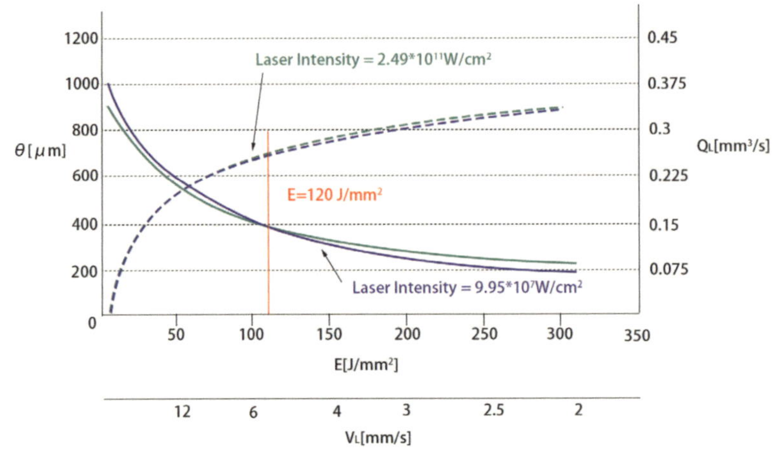

Figure 2.
Effect of laser input energy density per material removal rate. a_{e-L} *vs.* $E_{L-input}$ *and* Q_L *vs.* $E_{L-input}$. *Reproduced from ref. [19].*

on the direct interaction between the laser beam and the surface rather than the thermal diffusion in the material volume. Therefore, material ablation is mainly conducted in the direction of laser radiation. The energy distribution of the laser beam satisfies the Gaussian distribution model, with approximately 90% of the laser beam energy concentrated at the center of the laser beam (in spots with radii less than 30% of the laser spot radius) [22]. Reduce laser input energy density $E_{L-input}$ by increasing laser scanning speed can modulate the laser cutting depth at the cutting center, rather than the boundary of the irradiation area, to form a V-shaped laser cutting. In addition, laser ablation depth a_{e-L} for laser input energy density $E_{L-input}$ changes in a sensitive response compared to other cutting sizes. The correlation between laser input energy density $E_{L-input}$, laser ablation depth a_{e-L} and material removal rate Q'_L is shown in **Figure 3**. The relevant analytical relationship is shown in Eq. (1):

$$a_{e-L} = Bln\left(\frac{E_{L-input}}{A}\right) \tag{1}$$

In Eq. (1), A and B are constant values. Due to ablation depth a_{e-L} should always be zero or positive, the minimum value of laser input energy density $E_{L-input}$ should be equal to a. If $E_{L-input} = A$, then $a_{e-L} = 0.0$ μm. The values of A and B vary with the average pulse intensity I_L changes.

Eq. (1) is similar to the blow-off model for laser ablation, which assumes that exceeding characteristic threshold laser energy causes the ablation process. In this model, depth of ablation is given by the following equation [23]:

$$a_{e-L} = \gamma ln\left(\frac{E_{L-input}}{E_{th}}\right) \tag{2}$$

a_{e-L}: Laser ablation depth
E_{th}: Laser energy threshold
γ: Constant value.

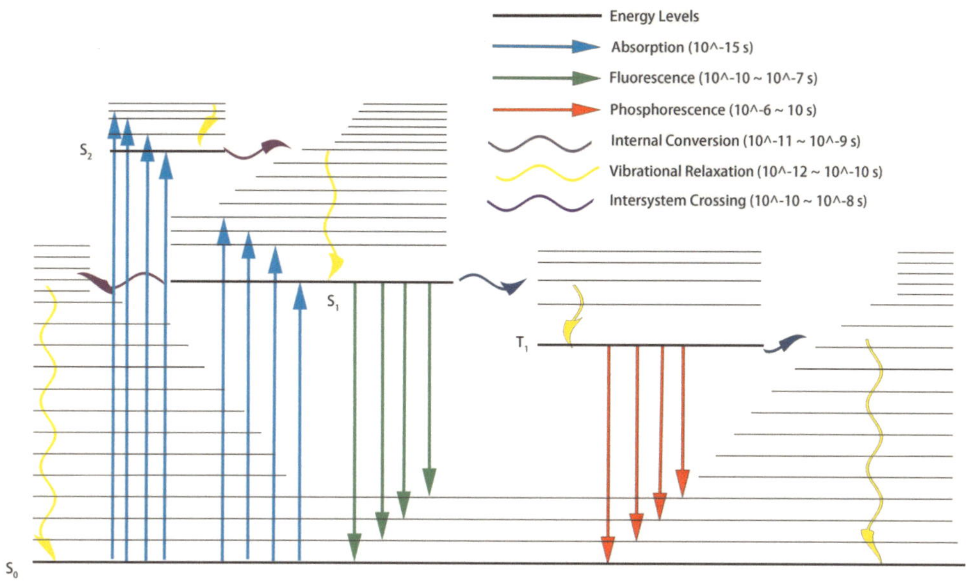

Figure 3.
Typical Jablonski diagram shows possible radiative and nonradiative transitions S_0 is the singlet ground state of the molecule, S_1 is the first excited singlet, S_n is the nth excited singlet, T_1 is the first excited triplet, and T_n the nth excited triplet.

The constants A and B in Eq. (1) depend on the inherent characteristics of the laser (such as wavelength λ and I_L) and material characteristics. In the blowing model, the decrease in I_L leads to the increase in threshold laser energy E_{th} and laser ablation depth and the decrease in ablation efficiency of each laser pulse [24]. The efficiency of laser ablation can be quantified by the process parameters, such as the specific ablation energy (e_L), as the laser energy required to ablate a unit volume of material. The lower the specific ablation energy e_L, the higher the laser ablation efficiency.

Based on nanosecond laser ablation, the interaction between laser and material is based on the thermal mechanism. From a thermodynamic perspective, material erosion is mainly due to the melting and evaporation of the irradiated material. The key ablation factor in the thermal process is the material's gasification process. During this process, if the energy absorbed by the molten or solid material is insufficient to evaporate, it only leads to heating and thermal damage to the irradiated surface. The consumption of laser energy in deep laser cutting leads to the decrease of I_L and results in an increase in material melting and thermal damage, rather than material evaporation.

Above, we have made a qualitative analysis and description of the interaction between laser and matter from the energy perspective. The underlying physical mechanism of laser-material interaction will be described and analyzed in the following sections.

3. Interaction of photons with semiconductor and dielectric

The mechanism that laser acts on the surface of semiconductor or dielectric materials to cause changes in the electronic state inside the materials can be explained by Jablonski diagram.

Figure 1 is a typical Jablonski diagram of the evolution process of state change and energy transfer of materials after light irradiation.

In the Jablonski diagram, several different pathways show how electrons receive and then dissipate the energy of photons from specific wavelengths. Therefore, most graphs begin with the electronic state of the ground state and then return to the ground state. The initial energy comes from the absorption of light interacting with matter, that is, the blue straight arrow in the figure. The absorbed energy will cause electrons to transition from the ground state to the excited state of a higher energy level. We focus on the physical connotation of absorption, vibrational relaxation and internal conversion, intersystem crossing, and timescale. The specific descriptions of the above cases are as follows:

3.1 Absorption

Absorption belongs to radiation transition because in this process, the molecule absorbs photons before it transitions from the lower electron level to the higher electron level. The energy of photons is converted into the internal energy of molecules. Absorption is the fastest transition in the Jablonski diagram, which occurs on the timescale of femtoseconds. At room temperature, based on the Boltzmann distribution, most molecules in the group will be in the lowest vibrational energy level of the ground state, and the absorption of photons will raise the molecule from S_0 to one of the vibrational energy levels of the single excited state ($S_1, S_2, \ldots$). Restricted by the conservation of angular momentum, the molecule cannot be directly excited to the triplet state excited state ($T_1, T_2, \ldots$).

3.2 Vibration relaxation and internal conversion

After being excited, electrons cannot be stably maintained in the excited state, but tend to return to the more stable ground state, which must be accompanied by energy dissipation. The first is through vibrational relaxation, a nonradiative process. This is represented on the Jablonski diagram as a curved arrow between the vibration levels. Vibrational relaxation is when the energy transferred from photons to electrons is fed into other vibrational energy levels as kinetic energy. This kinetic energy may remain in the same molecule or may be transferred to other molecules around the excited molecule. This process is also very fast, between and picoseconds. Since this is a very fast transition, and this relaxation occurs between the vibrational energy levels, usually electrons do not change from one electron level to another by this method.

However, if the vibrational energy level and the electronic energy level overlap severely, the excited electrons may transition from the vibrational energy level in one electronic state to another vibrational energy level in a lower electronic state. This process is called internal conversion and is similar in mechanism to vibrational relaxation. It is also represented on the Jablonski diagram as a curve between two vibration levels in different electronic states. The internal conversion occurs due to the overlap of the vibrational energy level and the electronic energy level state. This overlap increases the probability of electron transition between vibrational levels, which will reduce the electron energy level. Therefore, internal conversion and vibrational relaxation occur in the same time range, which is a way for molecules to dissipate energy from light excitation. However, due to the lack of vibration and the overlap of electronic energy states and the large energy difference between the ground state and the first excited state, the internal conversion of electrons back to the ground state is very slow. This slow return to the ground state causes other transfer processes to compete with the internal conversion in the first excited state. Both vibrational relaxation and internal conversion occur in most excitations, but rarely in the final conversion.

3.3 Intersystem crossing

Another way for molecules to absorb energy is through cross-mediation between systems. Based on the characteristics of electron spin multiplicity, we can relax from the excited singlet state to the excited triplet state. This is the slowest state transition process in the Jablonski diagram, several orders of magnitude slower than fluorescence. According to the transition selection rule, this slow transition was originally a prohibited transition. Based on the coupling of vibration factors and selection rules, transitions become weaker and can compete with the timescale of fluorescence. The crossover between systems results in different paths for returning the ground state electronic state. The direct transition is phosphorescence, which corresponds to the radiative transition from the excited triplet state to the singlet ground state. This is also a very slow and prohibited transition. Another possibility is to delay fluorescence and return to the first excited singlet state, causing the emission to change to the ground state electronic state.

Most molecules go from the excited state to the ground state through other nonradiative transition paths, so they do not exhibit fluorescence or phosphorescence behavior. On the one hand, they can transfer energy between molecules through molecular collisions or external transformations. On the other hand, energy is transferred between molecules through quenching, based on the overlap of absorption and fluorescence spectra. These pathways are all nonradiative processes, and when molecules relax to the ground state, they compete with fluorescence processes.

3.4 Timescale

The Jablonski diagram shows the types of transitions that may occur in specific molecules. The probability of each transition occurring depends on the timescale of the corresponding process, and the faster the process, the greater the probability of occurrence, based on the selection rules and their relaxation. **Table 1** lists the average timescales of basic radiative and nonradiative processes.

Transition	Timescale	Radiative process?
Internal Conversion	10^{-11}–10^{-9} s	No
Vibrational Relaxation	10^{-12}–10^{-10}s	No
Absorption	10^{-15} s	Yes
Phosphorescence	10^{-6}–10 s	Yes
Intersystem Crossing	10^{-10}–10^{-8} s	No
Fluorescence	10^{-10}–10^{-7} s	Yes

Table 1.
Average timescales for radiative and nonradiative processes.

It can be seen from the data in **Table 1** that almost all nonradiative energy transfer of excited electrons is completed in the nanosecond time range. This is the basis for the subsequent analysis of laser-material interaction.

4. Electronic mechanisms of semiconductor and insulator in ionization, desorption, and ablation induced by laser

4.1 Absorption and energy relaxation

In semiconductors and insulators, electron-hole pairs generated by photon absorption relax, forming localized and excited electron and vibration states. The electronic energy absorbed by photons is converted into nuclear motion and finally into desorption, which depends on the characteristics of the electron-lattice interaction. Of course, not all electron excitations lead to desorption. The electron-hole pair may lose energy or radiation recombination and be lost in the precursor channel for desorption. Various excitation and relaxation paths from the main photon solid interaction to the relaxation excited state are schematically shown in **Figure 1**.

The mass transfer and ionization mechanisms induced by laser can be roughly divided into thermal and nonthermal mechanisms. Thermal mechanisms include normal evaporation, photomechanical spalling, and phase explosion, while nonthermal mechanisms include electron transition-induced desorption, electrostatic ablation, Coulomb explosion, and so on. The peak power, tunable wavelength, continuous wave (CW), pulse duration, and precise beam optics of the laser beam provide micro/nano level controllable processing.

When a sufficiently strong laser pulse irradiates the material surface, the material undergoes different stages. These phases can be classified according to their duration. For example, chemical reactions occur on the femtosecond timescale (phonon vibration period). Similarly, electron motion is the key to any material modification, and it occurs on the timescale of attoseconds (the electrons need about 150 times per revolution of the atomic nucleus) [25]. This section explains these ultrafast mechanisms of electron state evolution and electron-lattice interaction in the nanosecond and femtosecond ranges. Furthermore, the desorption and ablation mechanism of the material under the action of a pulsed laser is explained.

The process of laser action on the material surface sequentially involves absorbing pulse energy, distributing energy to electronic subsystems, and then relaxing based on electron-phonon interactions [26, 27]. The absorption and distribution of energy occur within the femtosecond duration, while the electron-phonon interaction occurs within the picosecond duration. Such a nanosecond pulse interacts with matter, and all three processes will be completed within the duration of the pulse. The nanosecond pulse energy is used as a continuous energy source for the above three processes. When a laser with photon energy greater than Fermi energy is absorbed by the surface of the material, it directly leads to the excitation of electron transitions. The excited atoms relax through the combination of electron-phonon interaction, radiation recombination, Auger recombination, and carrier diffusion. According to the intensity of the laser pulse, the distribution is based on two different paths of heating and ablation, leading to melting. The excitation mechanism of ultrashort laser pulses causes the material to be in a strong nonequilibrium state characterized by hot electrons and cold ions. After the pulse ends, the electron transfers its energy to the ion through electron-phonon coupling in sub-picoseconds. This heats up the phonon bath before the slow thermal effect can reconstruct the material [18, 28–30]. In metals, electrons and phonons will reach the same temperature in this way, allowing for any subsequent thermal phase transition when the system is in internal equilibrium. In semiconductors and insulators, the generated excitons undergo additional recombination through radiative or nonradiative recombination, where the latter implies exchange with phonons [31]. Due to the presence of surface states and the formation of net charges after electron emission, the material surface needs to be regarded as an independent region with its own excitation and thermal mechanism. In other words, on the timescale equivalent to the relaxation time of the quasiparticle and the length scale similar to the average free path of quasiparticle, the quasiparticle pass through the structure unimpeded and mainly scatter at the interface. For example, in macroscopic objects, heat and phonon transport exhibit diffusion over a long period of time, but when the length and timescales become narrower, their properties become ballistic. The average free path of phonons can be as large as 300 nm but decreases with increasing temperature [31, 32].

In the case of femtosecond laser, a sufficiently strong laser pulse can generate free carrier density in the conduction band as high as 10^{20} to $10^{22}/cm^3$ [29, 33]. The high-density free charge carriers result in mutual shielding in the conduction band and heating before the relaxation of electron-phonon interactions. The time delay caused by free carrier shielding in the conduction band leads to bond weakening, causing band gap collapse, and leading to the transition from semiconductor to metal.

In addition, the electron-phonon relaxation process in femtosecond mode occurs after the pulse, and the duration of the femtosecond pulse is sufficient to cause the initial absorption part of the laser pulse energy to generate plasma on the material surface. The plasma absorbs the remaining part of the incident pulse through the reverse bremsstrahlung process (**Figure 4**). The absorbed pulse energy of the plasma accumulates deep in the material, resulting in a higher-density plasma [35].

4.2 Ionization mechanism

Multiple excitation mechanisms are used to explain the interaction between laser and condensed phases. Research has clearly pointed out the importance of laser and sample characteristics, as there is no universal mechanism to explain every situation. Starting from sufficiently high laser intensity for ablation or breakdown, the duration of the pulse determines whether thermal and hydrodynamic effects occur during the

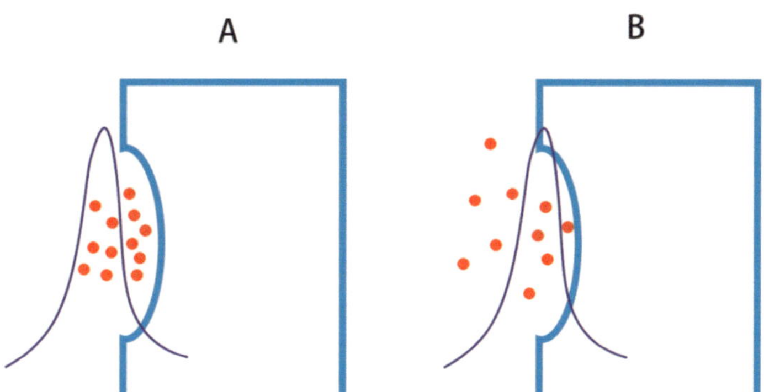

Figure 4.
Pulse propagation, (A) plasma generation, and (B) plasma expansion and absorption in femtosecond regime. Reproduced from ref. [34].

radiation period. For femtosecond pulses, this process can only begin after the irradiation is terminated. During nanosecond pulses, it is usually assumed that there is a balance between excitation and relaxation processes to maintain the material in an overall electrically unexcited state, and the energy of the transient excited electrons is rapidly transferred to the heavy species. This allows a purely thermal description using classical hydraulics and thermodynamics [6]. In the initial stage of laser interaction, nonthermal processes may have significant contributions that cannot be ignored. The wavelength of the laser further determines the properties of the excitation process. Ultraviolet radiation can usually cause electron excitation or ionization events using a single photon. In nonmetallic solids and liquids, ionization in this case corresponds to the transition of valence electron to the conduction band in the Jablonski diagram [36–39]. For a single photon with a longer wavelength, excitation and ionization can be realized through the nonlinear effect caused by high laser irradiance. Strong photon flow can be taken as a collective, and valence electron can be excited by field-induced ionization. Based on wave-particle duality, this mechanism can be distinguished as a function of multiphoton, tunneling, and barrier excitation intensity [40]. For example, multiphoton absorption enables infrared lasers to directly boost electrons within an energy range exceeding one of the constituent photons. Only when the ionization rate of the tunnel is fast enough relative to the frequency of light can quantum tunneling be achieved at higher irradiance [40]. From the moment of producing quasi-free electrons, excitation, and ionization can further occur through charge carrier acceleration and collision, as well as reverse bremsstrahlung radiation [4, 40]. Direct impact ionization should not be the main mechanism in dielectric materials. The above excitation mechanism can serve as a source for understanding the voltage-induced breakdown mechanism of condensed materials.

The dielectric surface is converted into plasma by ultrashort pulses of about 100 fs with an average intensity of about $1 \sim 10TW/cm^2$ [41, 42]. It is conventionally suggested that the ionization threshold is achieved when the electron number density in the conduction band reaches the critical density $n_c = m_e\omega^2/4\pi e^2$, corresponding to the incident laser wavelength [43]:

$$n_c = \int_0^{t_{ion}} W(I(r,z,t))dt \tag{3}$$

where $W\ (I\ (r,z,t))$ is the sum of ionization rates generated by electron collisions and multiphoton mechanisms, respectively. The ionization time is represented by the time required to reach the critical density. Therefore, the breakdown threshold depends on the injection amount. The degree of impact of collision and multiphoton ionization on the ionization rate depends on the oscillation energy of electrons in the laser field., $\varepsilon_{osc} = e^2E^2/4m_e\omega^2$ (averaged over the light period and for linear polarization), and the ionization potential, J_i, or band gap Δ_g [44]. The ratio of ionization potential (or band gap) to the energy of oscillations is the adiabaticity parameter, $\gamma = \Delta_g/\varepsilon_{osc}$. It can be used to separate different ionization states. When $\gamma > 1$, multiphoton ionization is dominant. Electron collision ionization is the main ionization mechanism in the long pulse region.

4.3 Desorption and ablation

The excitation and thermal processes in laser-irradiated materials lead to the rapid heating of phonon baths. For femtosecond laser pulses, rapid energy deposition leads to so-called inertial stress limitation [45, 46], which is due to the accumulation of strong compressive stress and the inability of materials to expand in such a short time. The subsequent relaxation of these stresses creates underground voids, leading to liquid surface layer separation and droplet ejection [45]. Therefore, during the ultrafast laser pulse, heat conduction and hydrodynamic movement are still limited, thereby reducing the thermal damage and heat-affected zone on the target [47]. There may be two explosion states, from the photomechanical spallation at lower flux to the phase explosion at higher flux [45]. Strong recoil pressures of up to 1*GPa* have also been reported for longer or continuous pulses [48], but the shielding of the laser beam by the jet plume can play a role, especially at longer wavelengths. This significantly reduces ablation efficiency compared to femtosecond pulses [49, 50]. However, for nanosecond pulses with shorter wavelengths, the ablation efficiency is still very high, and the inverse bremsstrahlung in the laser plume interaction can be ignored [4].

In this section, it is worth noting that three studies of Russo's group, one of which studied the particle size and shape [50], and two of which focused on the delayed particle emission effect [51, 52]. It was found that femtosecond pulses on brass produced particles with a diameter of about 100 nm, forming large aggregates, while nanosecond pulses produced spherical entities ranging from hundreds of nanometers to thousands of nanometers [50]. According to two other studies, a laser pulse on silicon with a duration of 3 ns first removes mass by normal evaporation during the pulse period and then sprays micron-sized particles due to delayed explosive boiling after 300 to 400 ns [51, 52]. This delay can be explained by the nucleation and growth delay of bubbles in the superheated melt. Note that this mechanism can also be effective for femtosecond pulses. Compared with nanosecond laser ablation, the limited heating of the material around the melt in the femtosecond case may even lead to no convex edge above the surface [53].

There are still many questions about the origin of molecular species in the ablation plume, because they may be by-products of evaporation or photomechanical spallation, or phase explosion, released directly from the sample surface or from ejected nanoparticles or droplets. In the case of femtosecond pulses, the generation of molecular fragments is high over a wide wavelength and flux range [7] and also depends on the position of the laser focus relative to the sample surface [54]. Nanosecond pulses are not suitable for maintaining molecular structure at infrared and ultraviolet wavelengths. For infrared, this is due to plasma heating caused by the interaction between low molecular generation rate at low flux and laser plumes at high flux [7]. Nanosecond ultraviolet pulses do not have this shielding effect and will destroy molecular bonds based on high photon energy and strong thermal effects. Two more controversial mass transfer mechanisms in laser ablation are electrostatic ablation and Coulomb explosion. Both assume that ions are directly released from the sample surface into the gas phase through a strong field effect. As a first step, the laser generates a positively charged surface layer through a sufficiently short wavelength photoelectric effect or field-assisted thermionic emission. According to the electrostatic ablation mechanism, surface ions are then pulled into the gas phase by the electric field induced by the emitted electrons [55, 56]. However, the Coulomb explosion is caused by the repulsive force between positive ions in the surface layer [55, 56]. The process occurs ultrafast, which explains the early ion emission during femtosecond pulse ablation (see **Figure 5**). This mechanism is also controversial, as it is believed that surface charges are quickly eliminated through charge compensation through the redistribution of conductive electrons within the body.

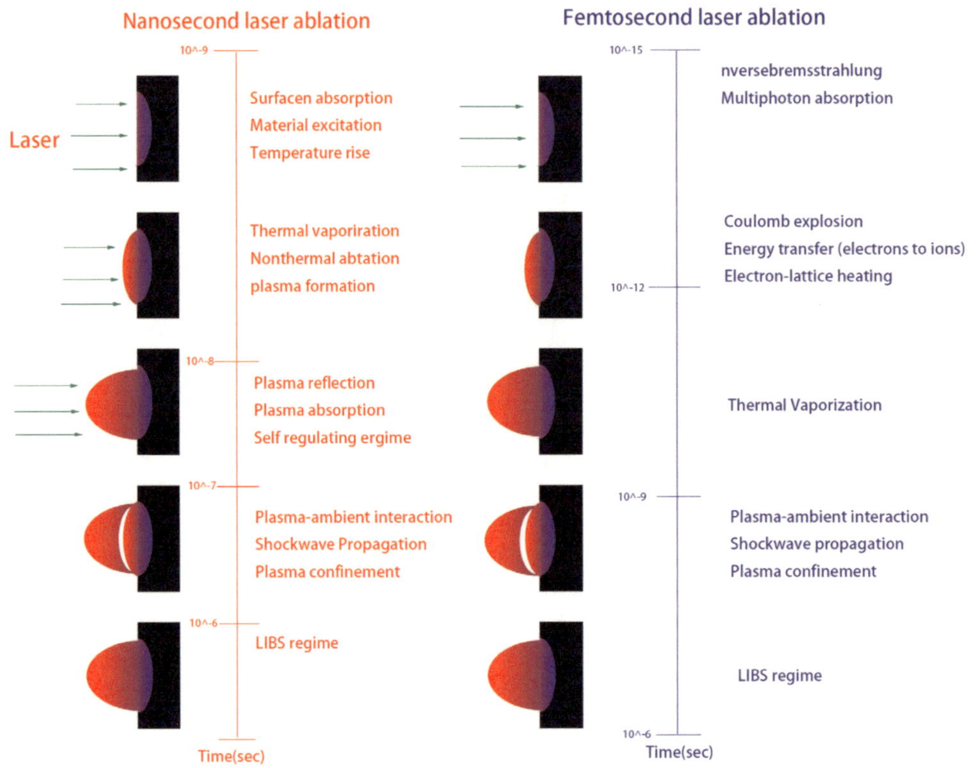

Figure 5.
Timescales of various processes taking place during and after the laser pulse in (left) nanosecond and (right) femtosecond laser ablation. Reproduced from ref. [4].

5. Electron-lattice interactions and material structural relaxation

5.1 Relaxation of a nonequilibrium phonon distribution induced by laser irradiation

Laser action on a substance first excites electrons, which transmit the stimulated energy to the nucleus and convert it into driving energy for nuclear motion, causing atoms, ions, and molecules to eject from the material surface, forming laser-induced desorption and ablation. Based on a series of electron lattice interactions, such as phonon scattering of free electrons, phonon emission, local lattice rearrangement, and configuration changes, and the rate and dynamic characteristics of transformation, are influenced. The configuration changes caused by the self-trapping of holes and excitons, defect formation and reaction, and electrical interactions between defects and lattice ions lead to surface decomposition.

Femtosecond laser excitation provides a suitable condition for studying the basic process in irradiated materials, because the duration of laser pulse is much shorter than the time required for the energy transfer process in the excited materials. **Figure 6** shows a comparison of thermal mechanisms, charge carrier removal, thermal effects, and structural effects. According to the ultrashort laser pulse excitation mechanism described in Section 4, laser radiation forms strong nonequilibrium states characterized by hot electrons and cold ions in the material. At the end of the pulse,

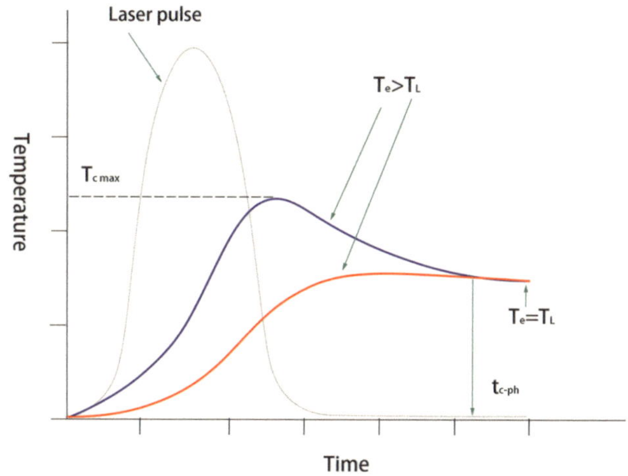

Figure 6.
Qualitative time dependence of electron and lattice temperature in the skin layer. The dotted line is a Gaussian shape of fs-laser pulse; t_{e-ph} indicates the energy equilibration time. Reproduced from ref. [56].

energy is transferred to the ion within sub-picoseconds through electron-phonon coupling. In semiconductors and insulators, excitons are generated simultaneously, and they will undergo radiative or nonradiative recombination, which involves exchanging energy with phonons. However, the simulation of silicon shows that the phonon system is in a nonequilibrium state within several hundred picoseconds [57]. Therefore, this nonequilibrium may affect the early stage of the thermal effect. In this regard, it seems to be more recommended to consider from the nonthermal and thermal timescales than from the nonthermal and thermal processes.

5.2 Dynamic mechanism of atomic site change in materials under the action of pulsed laser

In the initial stage of laser interaction with semiconductors and insulators, heated electrons are rapidly excited, and atoms vibrate almost undisturbed near their equilibrium positions. When the laser pulse ends, the electron reaches its maximum temperature and produces an electron temperature gradient in the surface layer, which generates a driving force on the surface layer. This force is proportional to the electron temperature gradient and acts on atoms, causing coherent motion of the atoms. This force is generated at a time shorter than the phonon period ($t_{ph} = 2\pi/\omega_{ph}$ ~few hundred *fs*) and produce rapid atomic shifts. The magnitude of this force is proportional to the absorbed energy density. The elastic force that causes atomic vibration in an undisturbed solid is also greater than the electric force at the energy density of the melting equilibrium enthalpy. The electric force only acts as a short pulse, triggering atomic vibration at the "cold" phonon frequency. From a time perspective, the next stage of atomic vibration corresponds to harmonic vibrations with "cold" phonon frequencies, which persist during the energy transfer from electrons to the lattice until the lattice reaches a temperature close to the melting point. Afterward, atomic vibrations gradually lose their harmonic characteristics, and nonlinear interactions between different phonon modes will dominate, ultimately transforming the solid into different phases or disordered states.

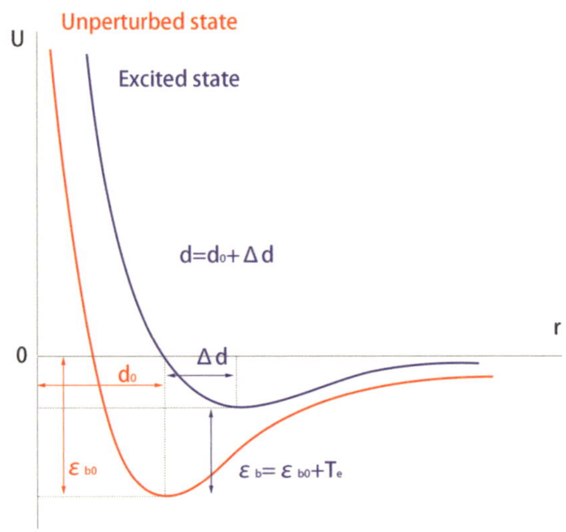

Figure 7.
Qualitative picture for the change of the potential energy due to electronic excitation.

The fast atomic motion in a fast excited solid can also be qualitatively described using the interatomic interaction potential (Morse potential) shown in **Figure 7**. In fact, the interatomic potential is composed of the part that attracts electrons and the part that repels ion nuclei. Fast electron excitation reduces the attraction and therefore the binding energy. The distance between atoms increases due to the excitation of electrons. Compared with the quantum harmonic oscillator model potential, the interatomic potential in Morse contains the nonbonded state that is absent from the harmonic oscillator model, so it can describe the nonharmonic effect, frequency doubling, and combination frequency [58]. Morse potential expresses the atomic displacement as a function of electron temperature, binding energy, equilibrium atomic spacing, and the gradient of attraction and repulsion of the potential.

Based on the consideration of the above model potential, analyze the atomic vibration in an undisturbed solid, consider first-order approximation, ignore spatial dispersion, and consider the excited phonon as a standing wave. The elastic force driving harmonic vibrations in solids can be represented by the second-order perturbation term in the interatomic potential, which is in the form of:

$$\Delta U_{el} = \frac{1}{2}\left(\frac{\partial^2 U}{\partial q^2}\right)_0 q^2 \approx \frac{1}{2} M\omega_0^2 q^2 \quad (4)$$

One estimates $\left(\partial^2 U/\partial q^2\right)_0 \approx \varepsilon_b/d^2$; here q is the cold phonon amplitude, ε_b is the binding(cohesive) energy, and d is the interatomic distance in equilibrium. The cold phonon frequency in Eq. (4) is $\omega_0^2 \approx \varepsilon_b/Md^2$. The elastic force driving harmonic vibrations immediately follows from Eq. (4):

$$F_{el} = \frac{\partial \Delta U_{el}}{\partial q} \approx M\omega_0^2 q \quad (5)$$

The cold phonon amplitude at a temperature lower than the Debye temperature is estimated as [59]:

$$q_0 \approx \left(\frac{2\hbar}{M\omega_0}\right)^{1/2} \tag{6}$$

Now, the elastic force explicitly expresses through the basic characteristics of a solid:

$$F_{el} \approx \left(M\omega_0^3\hbar\right)^{1/2} \tag{7}$$

Considering the effect of a laser electric field on the material under constant volume conditions, the laser causes internal atomic motion that leads to material deformation, thereby modulating the dielectric function of the medium. The stress tensor of the initial isotropic medium can be expressed as [60]:

$$\sigma_{ik} = -P\delta_{ik} - \left(\frac{E^2\delta_{ij}}{8\pi} + \frac{E_iE_j}{4\pi}\right)\varepsilon_{jk} + \frac{E^2}{8\pi}\left[n_a\left(\frac{\delta\varepsilon_{jk}}{\delta n_a}\right)_T\right]\delta_{ij} + \frac{E_iD_k}{4\pi}, \tag{8}$$

where the electric displacement vector has a form $D_k = \varepsilon_{kj}E_j$. We assume that the dielectric tensor modified by the laser effect consists of two terms, the Drude-like term, ε_D, and polarization term, $\varepsilon_{jk}^{(p)}$:

$$\varepsilon_{jk} = \varepsilon_D\delta_{jk} + \varepsilon_{jk}^{(p)} \tag{9}$$

The electron number density, dielectric function, electron and lattice temperature in transient are all time-dependent.

The volume force caused by the laser field can be represented by the stress tensor modulated by the field effect, σ_{ik}, taken from Eq. (8) as follows [61]:

The volume force caused by the laser field is expressed by the stress tensor σ_{ik} modified by the field action, which is taken from Eq. (8) as follows [61]:

$$f_i = \frac{\delta\sigma_{ik}}{\delta x_k} = -\frac{\delta P}{\delta x_k} + \frac{\delta\varepsilon_{ik}^{(p)}}{\delta x_k}\frac{E^2}{8\pi} + \frac{(\varepsilon_D - 1)}{8\pi}\frac{\partial E^2}{\partial x_i} = f_i^{th} + f_i^{(p)} + f_i^{pond} \tag{10}$$

Here, we consider that for the Druid-like part of the dielectric function, the following relation holds, $n_a(\partial\varepsilon_D/\partial n_a)T = \varepsilon_D - 1$. The first term on the right side of Eq. (10) is the thermal driving force caused by the above electron temperature gradient. The second term is the force related to polarization, based on the Plazcek effect [62], whose magnitude is inversely proportional to the atomic displacement, and the third term is the ponderomotive force. Polarization and ponderomotive force are only effective during the duration of the pulse, while thermal driving forces cause atomic movement until the temperature gradient is smooth throughout space. It is also worth noting that the polarization force in Eq. (10) is similar to but different from the force driving phonon excitation in the Raman effect [63]. This difference is related to the fact that the duration of the laser pulse is much shorter than the phonon period. Therefore, there is no interaction between the laser electric field and the vibration field of atomic motion. The phonon frequency enters the polarization force indefinitely.

The spatial variation based on optical power density forms the ponderomotive force acting on atoms or electrons or ions in the dielectric or plasma. The optical power density of a monochromatic plane wave is independent of distance, and the longitudinal component of ponderomotive force is zero. For traveling pulses in dispersive media, their spatial range varies with the variation of optical group velocity. At small group velocities, moderate pulse energy also transfers a large amount of momentum to the atoms colliding with it. The strong excitation of lattice atoms with large momentum at once leads to ionization, desorption, and ablation, which presents typical nonlinear characteristics.

Based on the above qualitative discussion about the structural relaxation of materials after pulse laser action. Briefly explain the ultrashort interaction mode, that is, the duration of the laser pulse is far less than the time of the energy equilibration in the irradiated material. Obviously, nonequilibrium phenomena dominate the material removal mode, one of which is the rapid formation of lattice distribution functions with truncated high-energy tails, which can lead to different ablation modes.

6. Conclusion

To sum up, this chapter first analyzes the interaction between laser and matter from the perspective of energy, describes the electronic response characteristics of semiconductors and insulating dielectric materials excited by nanosecond pulse laser and femtosecond pulse laser, explores the relaxation process of the interaction between excited electrons and lattice, and analyzes the generation mechanism of polarization force and pondermotive force acting on lattice based on the multiple coupling of laser field and plasma. At the same time, the basic analysis of the above laser-matter interaction process is given based on molecular dynamics. These effects may modulate the relevant physical processes of ionization, desorption, and ablation. It is hoped that the theoretical analysis results of the interaction between laser and materials in this chapter will have certain guiding significance for the related fields of laser processing, laser spectral analysis, and process design of periodic surface structure beyond the optical diffraction limit.

Author details

Hong Liu* and Chunlan Ma
School of Physical Science and Technology, Suzhou University of Science and Technology, Suzhou, China

*Address all correspondence to: hliu@usts.edu.cn

References

[1] Pariz I, Goel S, Nguyen DT, Buckeridge J, Zhou X. A critical review of the developments in molecular dynamics simulations to study femtosecond laser ablation. Materials Today: Proceedings. 2022;**64**: 1339-1348

[2] von Allmen M. Laser-beam interactions with materials. In: Springer Ser. Mater. Sci. Vol. 2. Berlin, Heidelberg: Springer; 1987

[3] Orazi L, Romoli L, Schmidt M, Li L. Ultrafast laser manufacturing: From physics to industrial applications. CIRP Annals - Manufacturing Technology. 2021;**70**:543-566

[4] Harilal S et al. Optical spectroscopy of laser produced plasmas for standard isotopic analysis. Applied Physics Reviews. 2018;**5**(2):021301

[5] Bogaerts A, Chen Z. Effect of laser parameters on laser ablation and laser-induced plasma formation: A numerical modeling investigation. Spectrochimica Acta Part B: Atomic Spectroscopy. 2005; **60**(9–10):1280-1307

[6] Bogaerts A et al. Laser ablation for analytical sampling: What can we learn from modeling? Spectrochimica Acta Part B: Atomic Spectroscopy. 2003;**58** (11):1867-1893

[7] Boueri M et al. Early stage expansion and time-resolved spectral emission of laser-induced plasma from polymer. Applied Surface Science. 2009;**255**(24): 9566-9571

[8] Yoo JH et al. Existence of phase explosion during laser ablation and its effects on inductively coupled plasma-mass spectroscopy. Analytical Chemistry. 2001;**73**(10):2288-2293

[9] Chen Z, Bogaerts A. Laser ablation of cu and plume expansion into 1 atm ambient gas. Journal of Applied Physics. 2005;**97**(6):063305

[10] Bogaerts A, Chen Z. Nanosecond laser ablation of cu: Modeling of the expansion in He background gas, and comparison with expansion in vacuum. Journal of Analytical Atomic Spectrometry. 2004;**19**(9):1169-1176

[11] Bonse J, Höhm S, Kirner SV, Rosenfeld A, Krüger J. Laser-induced periodic surface structures— A scientific evergreen. IEEE Journal of Selected Topics in Quantum Electronics. 2017; **23**(3):9000615

[12] Link S, El-Sayed MA. Optical properties and ultrafast dynamics of metallic nanocrystals. Annual Review of Physical Chemistry. 2003;**54**(1):331-366

[13] Hu M et al. Gold nanostructures: Engineering their plasmonic properties for biomedical applications. Chemical Society Reviews. 2006;**35**(11):1084-1094

[14] Zhang JZ, Noguez C. Plasmonic optical properties and applications of metal nanostructures. Plasmonics. 2008; **3**(4):127-150

[15] Wang L, Hasanzadeh Kafshgari M, Meunier M. Optical properties and applications of Plasmonic-metal nanoparticles. Advanced Functional Materials. 2020;**30**(51):2005400

[16] Linic S, Christopher P, Ingram DB. Plasmonic-metal nanostructures for efficient conversion of solar to chemical energy. Nature Materials. 2011;**10**(12): 911-921

[17] Taliercio T, Biagioni P. Semiconductor infrared plasmonics. Nanophotonics. 2019;**8**(6):949-990

[18] Henry H. Laser Precision Microfabrication. In: Sugioka K, Meunier M, Piqué A, editors. Berlin, Heidelberg: Springer; 2010. ISBN: 978-3-642-10522-7

[19] Soltani B, Azarhoushang B, Zahedi A. Laser ablation mechanism of silicon nitride with nanosecond and picosecond lasers. Optics and Laser Technology. 2019;**119**:105644

[20] Jandeleit J, Horn A, Weichenhain R, Kreutz EW, Poprawe R. Fundamental investigations of micromachining by nano- and picosecond laser radiation. Applied Surface Science. 1998;**127–129**: 885-891. DOI: 10.1016/S0169-4332(97) 00762-9

[21] Tuennermann A, Nolte S, Limpert J. Femtosecond vs picosecond laser material processing. Laser Technik Journal. 2010;**7**(1):34-38. DOI: 10.1002/ latj.201090006

[22] Kumar M, Melkote S, Lahoti G. Laser-assisted microgrinding of ceramics. CIRP Annals - Manufacturing Technology. 2011;**60**(1):367-370. DOI: 10.1016/j.cirp. 2011.03.121

[23] Vogel A, Venugopalan V. Mechanisms of pulsed laser ablation of biological tissues. Chemical Reviews. 2003;**103**(2):577-644. DOI: 10.1021/ cr010379n

[24] von der Linde D, Sokolowski-Tinten K. The physical mechanisms of short-pulse laser ablation. Applied Surface Science. 2000;**154–155**:1-10. DOI: 10.1016/S0169-4332(99)00440-7

[25] Kienberger E, Goulielmakis M, Uiberacker A, Baltuska V, Yakovlev F, Bammer A, et al. Atomic transient recorder. Nature. 2004;**427**(33): 817-821

[26] Steen WM, Mazumdar J. Laser Material Processing. Springer Verlag, fourth ed. 2010. ISBN 978-1-84996-061-8

[27] Li X, Guan Y. Theoretical fundamentals of short pulse laser–metal interaction: A review. Nanotechnology and Precision Engineering. 2020;**3**:105-125

[28] Rethfeld B et al. Modelling ultrafast laser ablation. Journal of Physics D: Applied Physics. 2017;**50**(19):193001

[29] Sundaram S, Mazur E. Inducing and probing non-thermal transitions in semiconductors using femtosecond laser pulses[J]. Nature Materials. 2002;**1**(4): 217-224. DOI: 10.1038/nmat767

[30] Yang G. Laser Ablation in Liquids: Principles and Applications in the Preparation of Nanomaterials. CRC Press; 2012. DOI: 10.1201/b11623. ISBN: 9780429086106

[31] Stolee JA et al. Laser–nanostructure interactions for ion production. Physical Chemistry Chemical Physics. 2012;**14** (24):8453-8471

[32] Joshi A, Majumdar A. Transient ballistic and diffusive phonon heat transport in thin films. Journal of Applied Physics. 1993;**74**(1):31-39

[33] Singh AP, Kapoor A, Tripathi KN. Ripples and grain formation in GaAs surfaces exposed to ultrashort laser pulses. Optics and Laser Technology. 2002;**34**:533-540

[34] Singh AP, Gaur SS. Attosecond laser-matter interaction: A short note. Materials Today: Proceedings. 2022;**64**: 295-297

[35] Maiuri M, Garavelli M, Cerullo G. Ultrafast Spectroscopy: State of the Art and Open Challenges. Journal of the

American Chemical Society. 2020; **142**:3-15

[36] Lazic V, Jovićević S. Laser induced breakdown spectroscopy inside liquids: Processes and analytical aspects. Spectrochimica Acta Part B: Atomic Spectroscopy. 2014;**101**: 288-311

[37] Dharmadhikari J et al. Optical control of filamentation-induced damage to DNA by intense, ultrashort, near-infrared laser pulses. Scientific Reports. 2016;**6**(1):1-9

[38] Linz N et al. Wavelength dependence of femtosecond laser-induced breakdown in water and implications for laser surgery. Physical Review B. 2016;**94**(2):024113

[39] Vanraes P, Bogaerts A. Plasma physics of liquids - a focused review. Applied Physics Reviews. 2018;**5**(3): 031103

[40] Balling P, Schou J. Femtosecond-laser ablation dynamics of dielectrics: Basics and applications for thin films. Reports on Progress in Physics. 2013;**76** (3):036502

[41] Du D, Liu X, Korn G, Squier J, Mourou G. Laser induced breakdown by impact ionization in SiO_2 with pulse widths from 7 ns to 150 fs. Applied Physics Letters. 1994;**64**:3071-3073

[42] Stuart BC, Feit MD, Herman S, Rubenchik AM, Shore BW, Perry MD. Optical ablation by high-power short-pulse lasers. Journal of the Optical Society of America B. 1996;**13**: 459-468

[43] Kruer WL. The Physics of Laser Plasma Interactions. New-York: Addison-Wesley Publishing; 1988

[44] Gamaly EG, Rode AV, Luther-Davies B, Tikhonchuk VT. Ablation of solids by femtosecond lasers: Ablation mechanism and ablation thresholds for metals and dielectrics. Physics of Plasmas. 2002;**9**:949-957

[45] Kramer T et al. Ablation dynamics–from absorption to heat accumulation/ ultra-fast laser matter interaction. Advanced Optical Technologies. 2018; **7**(3):129-144

[46] Zhigilei LV, Lin Z, Ivanov DS. Atomistic modeling of short pulse laser ablation of metals: Connections between melting, spallation, and phase explosion. Journal of Physical Chemistry C. 2009; **113**(27):11892-11906

[47] Chichkov BN et al. Femtosecond, picosecond and nanosecond laser ablation of solids. Applied Physics A. 1996;**63**(2):109-115

[48] Yilbas B et al. Laser-shock processing of steel. Journal of Materials Processing Technology. 2003;**135**(1):6-17

[49] Zeng X et al. Experimental investigation of ablation efficiency and plasma expansion during femtosecond and nanosecond laser ablation of silicon. Applied Physics A. 2005;**80**(2):237-241

[50] Liu C et al. Nanosecond and femtosecond laser ablation of brass: Particulate and ICPMS measurements. Analytical Chemistry. 2004;**76**(2): 379-383

[51] Lu Q et al. Delayed phase explosion during high-power nanosecond laser ablation of silicon. Applied Physics Letters. 2002;**80**(17):3072-3074

[52] Lu Q et al. Theory analysis of wavelength dependence of laser-induced phase explosion of silicon. Journal of Applied Physics. 2008;**104**(8):083301

[53] Russo RE et al. Femtosecond laser ablation ICP-MS. Journal of Analytical Atomic Spectrometry. 2002;**17**(9): 1072-1075

[54] Lee Y et al. Spatial and temporal distribution of metal atoms and their diatomic oxide molecules in femtosecond laser-induced plasmas. Journal of Analytical Atomic Spectrometry. 2018;**33**(11): 1875-1883

[55] Ionin AA, Kudryashov SI, Samokhin AA. Material surface ablation produced by ultrashort laser pulses. Physics-Uspekhi. 2017;**60**(2):149

[56] Gamaly EG, Rode AV. Physics of ultra-short laser interaction with matter: From phonon excitation to ultimate transformations. Progress in Quantum Electronics. 2013;**37**(5):215-323

[57] Klett I, Rethfeld B. Relaxation of a nonequilibrium phonon distribution induced by femtosecond laser irradiation. Physical Review B. 2018;**98** (14):144306

[58] Morse PM. Diatomic molecules according to the wave mechanics. II. Vibrational levels. Physical Review. 1929;**34**:57

[59] Il'inskii YA, Keldysh LV. Electromagnetic Response of Material Media. New York: Springer; 1994

[60] Landau LD, Lifshitz EM, Pitaevskii LP. Electrodynamics of Continuous Media. Oxford: Pergamon Press; 1984

[61] Kent AJ, Kini RN, Stanton NM, Henini M, Glavin BA, Kochelap VA, et al. Acoustic phonon emission from a weakly coupled superlattice under vertical electron transport: Observation of phonon resonance. Physical Review Letters. 2006;**96**:215504

[62] Gamaly EG. The physics of ultra-short laser interaction with solids at non-relativistic intensities. Physics Reports. 2011;**508**:91-243. DOI: 10.1016/j.physrep.2011.07.002

[63] Shen YR, Bloembergen N. Theory of stimulated Brillouin and Raman scattering. Physical Review. 1965;**137**: 1787-1805

Chapter 2

Pulsed Laser Ablation in Liquids for Fabrication of Noble Metal Nanostructures

Oana Andreea Lazar, Anastas Savov Nikolov, Călin Constantin Moise and Marius Enachescu

Abstract

Pulsed laser ablation in liquids (PLAL) is a physical method that is applied for the fabrication of the noble metal nanostructures with different optical and morphological properties. The physical mechanism of laser ablation in a liquid environment, the subsequent growth of nanostructures, the essential laser technological parameters that determine the nanostructures' properties, and the liquid medium's influence are discussed. The main advantages and disadvantages of the PLAL method are noted. Post-ablation treatment at the optimal laser wavelength, fluence, and duration of exposure has been indicated as a means of mitigating and overcoming the latter. The aging effect of the colloids and some applications of them are also marked. The most commonly used methods for studying the nanostructures' characteristics such as UV/Vis spectroscopy, high-resolution scanning transmission electron microscopy (HR-STEM), mass spectrometry (MS), and X-ray diffraction (XRD) are commented.

Keywords: PLAL, nanoparticles, UV/Vis, HR-STEM, MS, XRD

1. Introduction

The first ruby laser was developed by Maiman nearly 60 years ago in Hughes Research Laboratories and was employed to irradiate and to ablate materials [1–3]. After its appearance, the pulsed laser ablation (PLA) method was used to produce various types of nanostructures such as thin films (TFs), nanoparticles (NPs), nanowires (NWs), or nanonetworks (NNs), which may be further involved in distinguished applications [4–7]. NPs (particles with dimensions in the interval from 1 nm to 100 nm) have attained significant interest in the past and even more nowadays because of their unique physicochemical properties (optical properties, electrical conductivity, melting point, density, chemical stability, etc.), which differ from that of the corresponding bulk materials [8–12]. Their specific characteristic is the large number of surface atoms compared to the number of such atoms in bulk, associated with high reactivity and determining their catalytic, optical, and magnetic properties. Another particularly significant feature is the strong dependence of these properties on the shape and size of nanomaterials. The latter provides an opportunity to create nanomaterials with independently and precisely controlled properties, such as size, composition, morphology, defect density, and atomic structure, that meet the specific requirements of a distinct application. Based on the material they are made of, nanostructures can be classified into four separate groups containing fullerenes, metallic NPs, ceramic NPs, and polymeric NPs, respectively [13–16].

Due to their optical properties, the existence of surface plasmon resonance (SPR) resulting from an interaction between electromagnetic waves and electrons in the conduction band (plasmonic process) and the ability to control the optical field, metallic NPs are of great interest in nanotechnology, whose subject is the different fabrication strategies and morphological modifications of nanomaterials. Plasmonic materials are well-known for their harsh interaction with free electrons and incoming photons. Thus, metallic materials can function as a source in order to convert the light into a local electric field in metals, named surface-localized plasmonic [17–20]. These interactions can also be tuned by modifying the morphological properties and the diameter size of the metal nanostructures. A special class of metallic NPs is represented by the noble metal nanoparticles (NMNPs), such as Ag, Au, Pd and Pt, characterized by tunable optical and remarkable photoelectric properties, high corrosion and oxidation stability, and exhibit a good biocompatibility related to their low biotoxicity. Based on these properties, an amazing variety of fields of application is established. We will mention some of them for illustration: in nano-enhanced catalysis and electrocatalysis, new energy materials, photoelectric information storage, for stable dyeing in the textile industry, colorimetric, and fluorometric sensors, diagnostic sensors in micro and nanoelectronics, for micro-joining of several electronic components in microelectronics, microelectronic systems for screen printing and inkjet printing, for drinking water purification decomposition of hazardous pollutants in the environment to minimize environmental pollution, bio diagnostics and biology, labels or probes in biosystems, as a biomaterial because of their anti-microbial, anti-inflammatory, anti-sterile and anti-allergic capability, in photothermal therapy, diagnostic and imaging applications, cancer treatment and diagnostic, for targeted delivery of active agents against tumor cells, etc. [21–31].

NMNPs could enhance the response of organic or biomedical material having the plasmonic influence in the ultraviolet-visible spectrum (UV-Vis) because of the photon scattering and absorption. With their widespread use, various methods have been developed for the synthesis of metal, in particular, NMNPs. They can be classified into two main groups—top-down and bottom-up approaches. Applying physical and chemical techniques to the bulk materials to reduce them to nanosized particles is the essence of the first procedure. The essential disadvantages are large size variation, imperfect surface structure and expensive processing equipment to maintain high pressure and high-temperature conditions during synthesis. In the bottom-up procedure, nanoparticles are built by assembling atoms, molecules or clusters through chemical and biological processes. This method provides better control over the forming process, a more homogeneous size distribution, and chemical composition. It is generally less expensive but is limited by the use of toxic chemicals, organic solvents and reagents [32].

Another classification is also possible according to the processes involved in the manufacturing methods, specifically physical, chemical or biological [32–41]. In general, regardless of the specifics of each of them, they present the disadvantages listed.

Synthesis of nanomaterial structures by pulsed laser ablation (PLA), which is a top-down physical method, can be performed either in a high vacuum deposition chamber at different pressure values with or without gases or in a liquid medium. The implementation of PLA in a liquid environment (PLAL), also known as laser ablation synthesis in solution (LASiS), is an approach that overcomes the main shortcomings of the other methods. It does not require expensive equipment; its application is under ambient pressure and temperature conditions and can produce chemically pure nanoparticles free of potentially toxic impurities. This approach was first reported in 1987 by Patil et al., and since then and especially in recent times, it has been frequently used [42]. In addition, the PLAL process can be split into various other methods to create complex nanomaterials including reactive PLAL, electric field-assisted PLAL, magnetic field-assisted PLAL, electrochemical PLAL and sequential PLAL [43–47]. Through this method, distinguish nanomaterials can be produced, such as pure metals, metal alloys, metal oxides, sulfides, polymer NPs, organic materials, semiconductors, ceramic nanostructures, nanocomposites with many different morphologies including core-shell, nanocubes (NCs), nanorods (NRs), NWs, NNs and other complex composites as a nano-resource for optical/photonics, catalytic, energy and biological applications [14, 15].

2. Experimental setup

The PLAL method used for the synthesis of nanostructures has a simple and basic setup in which three essential elements such as a laser, a target and a liquid medium are required. In other words, it consists of a laser system (ms, ns, ps or fs) [48–52], an optical system and a container where the target is placed. **Figure 1** shows the most used experimental setup for the PLAL method. Among the systems mentioned above, at least one of them is more expensive, and that is the laser system, but if you have a laser in the lab that can be used, then the other elements are less expensive. Almost all literature has reported the use of an identical diagram, except for systems using a rotating table or magnetic

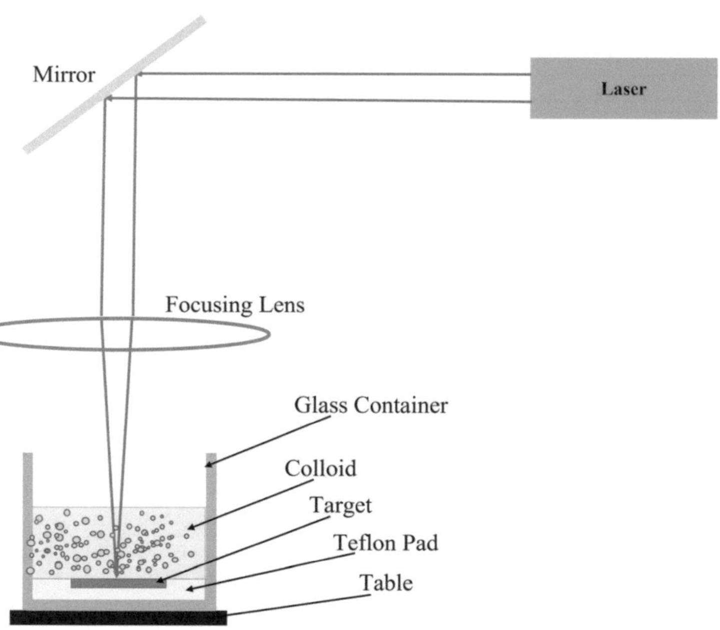

Figure 1.
Experimental setup of PLAL process.

stirring of a liquid. Therefore, the target is placed on the bottom of the container, the latter is located on a movable XY stage, and the optical system is adjusted so that the laser beam should reach the target. Generally, the target is located prior to the focal plane of the optical system, specifically, the focusing lense, to avoid optical breakdown in the liquid medium during the ablation process, while the laser spots on the target surface have diameter sizes ranging from millimeters to microns. Also, suitable modifications to the experimental setup can be made in order to increase the efficiency of the ablation rate by positioning the target at different angles (0°, 45°, 90°) during PLAL processes [53–57].

As mentioned above, the technological parameters of the PLAL such as pulse duration, laser wavelength, laser fluence, laser energy, repetition rate (RR), number of shots, ablation time, the nature of target, solvent and solutes have an important role in obtaining NPs with different optical and morphological properties, but the laser parameters are fundamental and required when PLAL occurs. Typically, laser systems with different pulse durations from nanoseconds up to femtoseconds operating from ultraviolet to infrared wavelengths were utilized in the synthesis of NPs applying a broad range of laser fluence values [52]. It should be noted that the usages of PLAL in industrial applications have some limitations regarding the difficulties to control the mean size (MS), standard deviation (SD) and size distribution, respectively, for NPs and their production rate if mass concentration is lower in a shorter ablation time. Among the advantages is that the laser ablation process only needs a minimal working procedure in which you will learn how easily the NPs can be created, the bulk material to be ablated instead of metal salts or chemical reagents, and minimal waste fabrication during PLAL instead of getting toxic waste by using other synthesis methods.

In the following sections, the technological parameters and the synthesis mechanism of growing NPs, as mentioned above, will be described and commented in detail for a better understanding of when the PLAL process is used.

3. Basic technological parameters

Fundamentally, the main technological parameters such as laser wavelength, laser beam width, laser fluence, laser RR, laser ablation duration, and desired medium are involved during PLAL process in the formation of the NPs.

The laser wavelength (λ) is an important parameter when MNPs with the desired morphology are produced through the laser ablation process. Using different types of laser sources, from Nd:YAG lasers to excimer lasers (KrF, ArF, XeCl), NPs with different shapes or sizes can be obtained [58]. As known from the literature, Nd:YAG laser is mainly used in PLAL process to create NMNPs as a laser source with a fundamental wavelength of 1064 nm because of their simplicity and capability to double, triple, or quadruple the frequency by optical systems [59]. Excimer lasers such as ArF (λ = 193 nm), KrF (λ = 248 nm) or XeCl (λ = 308 nm) could deliver short pulses of tens of nanoseconds and, due to their versatility of ultraviolet light, may be a choice in the laser ablation procedure to ablates different and numerous materials [60]. Mafune *et al.* utilized the fundamental wavelength (λ = 1064 nm) and second harmonic (λ = 532 nm) in pure water to prepare Pt-NPs with an average diameter of 6.2 nm [61]. For instance, Mortazavi *et al.* fabricated Pd-NPs using two different laser sources, Nd:YAG (1064 nm) and ArF (193 nm), and their influence on the production and structure of the obtained NPs was studied [62]. Censabella *et al.* produced Pt-NPs with a mean diameter size of around 10 nm also by an Nd:YAG laser in deionized water [63]. Nikolov *et al.* utilized the fundamental (λ = 1064 nm), second

(λ_{SHG} = 532 nm), third (λ_{THG} = 355 nm), and fourth harmonic (λ_{FHG} = 266 nm) of a Nd:YAG laser system to ablate Ag-NPs. The best result was achieved with the fundamental wavelength, specifically, an average size of 12 nm and a standard deviation of 4 nm [64].

The laser beam width (τ). The working principle of the PLAL process depends on the pulse duration, which represents the time at which the laser energy is driven to the target surface and as a result, different effects such as light absorption, heat generation and target ionization can occur. Various scientific papers have reported that the pulse width probably significantly influences the structure and size of NMNPs. The effect of pulse width on the growth mechanism for the fabrication of NMNPs in the initial phase of PLAL was investigated by Shih *et al.* [65]. The influence of the femtosecond pulses on Au-NPs synthesized in water was studied by Kabashin *et al.* [66].

Moreover, *laser fluence* is an essential laser parameter that can influence the PLAL process during the NMNPs formation and evaluates the quantity of laser energy per area on the ablated target (J cm^{-2}). The minimum laser energy required to ablate the target and achieve vaporization during the PLAL procedure is also called threshold laser fluence. The latter depends on the type of pulses used during the ablation process (i.e., ns, ps, or fs pulses) and can have different behaviors such as being independent of the pulse length (for ns pulses) or being dependent on the pulse length (for ps or fs pulses). Several scientific papers have been reported in the study of the influence of laser fluence during the formation of NMNPs. Amendola *et al.* produced Au-NPs in water and dimethyl sulfoxide solution using laser fluence values between 12 and 442 mJ cm^{-2} [67]. By using these values, a trend of increasing and decreasing NMNPs diameter sizes can occur. A theoretical explanation of the dependence Au-NPs diameter size on the laser fluence used to ablate Au target was made by Pyatenko *et al.*, while this correlation was studied experimentally by Tsuji *et al.* [68, 69]. The impact of the laser fluence on the Ag-NP characteristics ablated in distilled water was investigated by Xu *et al.* [70]. They observed that the MS values of Ag-NP diameter decrease when the applied laser fluence is less than 6.4 J cm^{-2} and increases when its value is higher. They established a decrease in size distribution when the fluence is less than 4.2 J cm^{-2} and higher than 6.4 J cm^{-2} and increases between these two values. This result matches the full width at the half maximum (FWHM) of the corresponding plasmonic band in the UV/Vis absorption spectra. Regarding the average NPs diameter size, it initially decreases from about 22 nm to about 17 nm when the laser fluence changes from 3.4 to 4.2 J cm^{-2}. With a subsequent increase of the laser fluence to 8.4 J cm^{-2}, the average value of the diameter increases, reaching about 39 nm at the indicated fluence. Moniri *et al.* fabricated Pt-NPs with a mean size value of 18 nm by using the fundamental wavelength at a RR of 10 Hz and laser fluence of 2 J cm^{-2} [71]. Lazar *et al.* fabricated Pt-NPs in double-distilled water (DDW) using KrF excimer laser by varying the laser fluence (2.3, 4.0 and 5.8 J cm^{-2}) and RR (10, 20, 30 ,40 and 50 Hz) while the time ablation was fixed at 10 min. They observed that by increasing the laser fluence and keeping the RRs constant, the mean size enhanced from 2.2 nm up to 4.0 nm, while the smallest mean size value of 2.2 nm for Pt-NPs created with the smallest laser fluence was achieved [72].

Pulse repetition rate (RR) is a fundamental laser parameter and depends on the applied laser energy used in the PLAL process. As it is known from other scientific works, RR is defined as the number of pulses released per second. Moreover, RR optimization and adjustment are necessary because of the shielding effect that can be generated by the plasma over laser pulses and can influence the yield of the created NPs. Therefore, to minimize this effect, the RR variation, the laser spot size and a suitable movement between the laser beam and target are needed. The impact of the RR varying from 1 Hz to 10 Hz in the productivity of ablated Ag-NPs in ethanol was studied by Valverde-Alva *et al.* [73]. They observed that the laser ablation efficiency of NPs grew with the enhancement of the laser RR. Pt-NPs ablated in aqueous ethanol solution using KrF excimer laser were obtained by Lazar *et al.* [74]. By varying the laser RRs from 10 Hz to 50 Hz using 2.3 J cm^{-2} laser fluence in 40% ethanol solution, an enhancement in the mean size values from 2.0 nm up to 2.45 nm was observed, while the standard deviation values remained similar at ± 0.8 nm. Altowyan *et al.* produced bimetallic NPs from Ag and Au, and they studied the effect of the laser fluence on the created nanostructures [75]. They also showed that the formation of Au (core)-Ag (shell) depends on the applied laser energy, and the shell thickness enhances with increasing laser energy. The influence of the laser fluence on the properties of the fabricated Pd-NPs was investigated by Mendivil *et al.* [76]. The formation of cubic nanostructures was observed, the MS and SD values simultaneously decreased from 27 ± 9 nm to 17 ± 6 nm with increasing laser fluence varying from 8 to 40.5 J cm^{-2}.

In general, the target material and the liquid environment used in the pulsed laser ablation process are also among the most important technological parameters. The optical and morphological properties of the obtained colloids depend on the material type used as the target and on the liquid medium employed to ablate the target.

Usually, the target that is utilized in PLAL method to obtain different nanostructures could be made from various materials such as pure metals (i.e., Ag, Au, Pt, Pd and Cu), metal alloys (i.e., AgAu, PtPd, FePt, AuFe, AgFe, PtCu and PtAu), carbon-based targets (graphite and graphene oxide), oxides (GaO and SiO), selenides (ZnSe and CdSe), nitrides (GaN and TiN), sulfides (PbS and CdS), tellurides (CdTe), antimonides (InSb), and pressed targets (Cu/Zn and WO_3/GO) [63, 77–96].

Normally, the properties of the liquid environment, such as height level over the target surface, the relative movement and the physical/chemical nature, can affect the nanostructures properties and the efficiency of the laser ablation process [97–99]. The height level of the liquid on the top of the target surface could influence the efficiency of the laser ablation process. Two groups, Jiang *et al.* and Al-Mamun *et al.*, showed that when a lower height level of liquid of 1.2 mm and 2.0 mm was used, a higher ablation rate was achieved [97, 100]. They affirmed that a confinement effect could occur at a low height level of the liquid above the target surface, and a high efficiency can be obtained as a result. Cristoforetti *et al.* ablated Pd-NPs in different liquids media such as water, acetone, ethanol, 2-propanol, toluene and n-hexane. They observed a lower efficiency in ablation process when water, acetone, toluene and n-hexane were used, while in ethanol and 2-propanol a comparable efficiency was achieved [101]. Moura *et al.* synthesized Ag-NPs in DDW, acetone and ethanol. The smallest NPs were achieved in DDW and the biggest in acetone and ethanol [102]. Lee *et al.* fabricate Au-NPs in water, methanol, hexane and acetonitrile, obtaining spherical, polycrystalline nanostructures and agglomerated chains. They also observed an enhancement in the catalytic activity of Au-NPs [103].

Additionally, the composition of the liquid and their additives can be classified into different batches like water and hydrogen-peroxide, acids and alkalis, organic solvents, monomers and polymers, salts, surfactants, superfluid and supercritical fluids, and combined liquids [15].

4. Ablation mechanism and nanostructures formation

The working mechanism of the PLAL remains misunderstood due to the physico-chemical process including laser-matter interaction (electron excitation, lattice heating and lattice disintegration), plasma plume, cavitation bubble, and liquid phase, which are involved during the NMNPs formation [104].

In essence, the PLAL process is based on the removal of the matter from the target through a laser beam that irradiates the target surface submerged in a liquid environment. The interaction between the laser beam and the target material and resulting ablation depends on the pulse duration, the liquid medium, the sample properties, and the focusing conditions used during the PLAL procedure. Usually, when a pulsed laser beam hits a metal plate, various materials can be carried out from the target surface in the form of vapors, fragments or clusters, which also interact with the liquid environment in order to form NMNPs [105]. Frequently, when short-pulse lasers with high laser fluence are used, the mostly vapors are produced, while using long-pulse lasers with lower laser fluence, mainly fragments or nanoclusters are created. Two processes have been identified in the literature, such as thermal vaporization of species from the liquid-solid interface and thermally induced emission of clusters/fragments from the surface target [1].

Supplementary, several steps during the ablation process were suggested like the easiest formation of atoms in the metal plate was realized by laser fluence from the laser beam. In the meantime, a plasma plume with high temperature was produced at the interface between the target and the liquid. Thus, the concentrated species within the plasma plume can be fast thermally scattered, resulting from their numerous aggregations, collisions and condensations. Lastly, as the plasma plume was extinguished by the surrounding liquid, the temperature of the plume began to decrease, and the obtaining nuclei started to spread and agglomerated into a larger nanostructure [105]. Nevertheless, the expansion of the plasma plume is confined by the surrounding liquid in order to form bubbles consisting of ablated species in gaseous form, which can lead the laser-induced plasma into thermodynamic conditions with high temperatures varying from 4000 K to 6000 K, high- density and high-pressure [106–109]. This thermodynamic condition of the plasma is interconnected with the laser parameters and the lifetime of the bubbles [107, 110, 111]. After that, the plasma plume will extend into the ambient liquid. During this expansion, the rapid pressure gradient caused by the shock wave emission and the energy transfer from plasma to liquid creates the cavitation bubble. Thus, during the laser ablation process, since the cavitation bubble can extend and reduce with time ablation, the high-pressure condition when the cavitation bubble drops down can impact the formation of the NMNPs [112]. The representative schematic of the laser-induced cavitation bubble can also be seen in **Figure 2**.

Due to the falling of bubbles, the nanostructures are created during the cooling stage of the plasma plume, and afterward, the NMNPs will be scattered in the surrounding liquid to compose the colloidal solution. In addition, during this procedure, the bubbles release huge energy into the liquid with a considerably minimized temperature and pressure, which determines the rapid extinguishment of the plasma and the emission of the ablated material [113–115]. Furthermore, a representative illustration can be drawn that consists of the fundamental steps described above about the ablation mechanism and this can be observed in **Figure 3**.

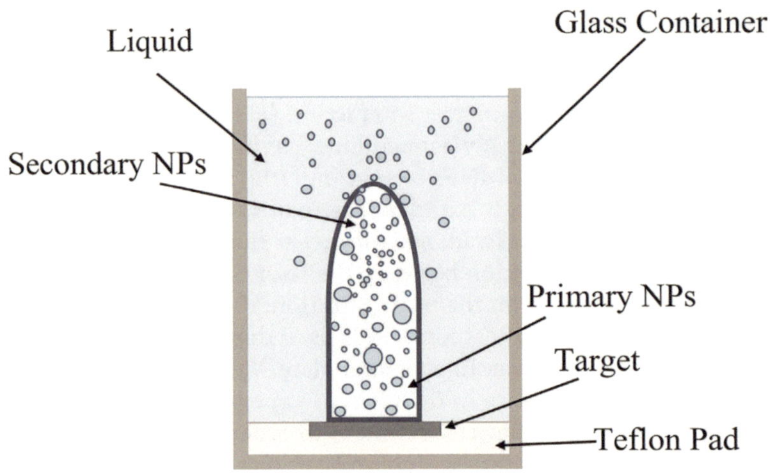

Figure 2.
Scheme for the cavitation bubble formation.

Figure 3.
Experimental steps that occur when the ablation mechanism begins.

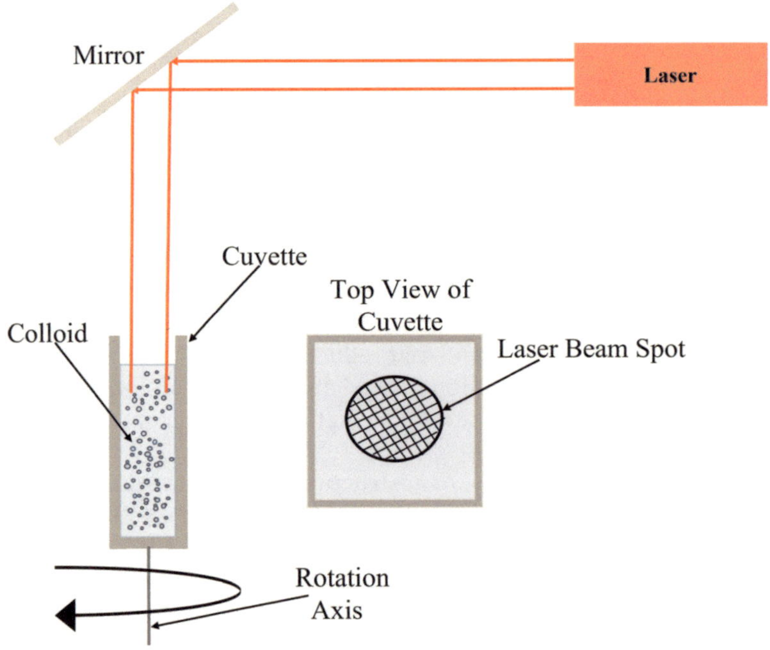

Figure 4.
Representative scheme of the laser fragmentation in liquid.

As it is already well-known, the synthesis of the nanostructures in the laser ablation field can be carried out either in the non-aqueous medium or in the liquid environment. The adjustable laser synthesis method of nanomaterials in a liquid environment has been widely used to create diverse colloidal nanostructures. Three laser synthesis techniques—laser ablation in liquids (LAL), laser fragmentation in liquids (LFL), and laser melting in liquids (LML)—can be distinguished based on the processing mode and on the formation mechanism of NMNPs [23]. Then, in this section, we reviewed the most current research in the nanomaterials field that utilized new LAL and LFL liquid phases. **Figure 4** shows the representative illustration of the experimental setup for an LFL technique type.

Furthermore, the liquid environment used in the PLAL can be non-reactive or reactive. By using a non-reactive liquid medium, the material ejected from the target will not interact with the liquid compounds, and the final colloidal solution will contain only the material from the target. When the nanostructure synthesis takes place with a reactive liquid medium, the atomic species from the target will interact with the molecules from the surrounding liquid leading to the new nanostructures [116].

5. Methods to study the PLAL process

PLAL is a complicated method whose implementation is related to the occurrence of physical and chemical processes under non-equilibrium conditions in order of nano-, pico-, and femtoseconds time scale. Their study must be carried out by ensuring sufficient temporal and spatial resolution. Such processes are the laser-matter interaction when the laser pulses are short or ultrashort, plasma formation, cavitation bubbles appearance and their dynamics, and the nanoparticles creation. This investigation, in addition to purely theoretical interest, also aims to manage them in order to control the characteristics of NMNPs as a final product. Several extensive review articles have been devoted to this topic [109–112].

Only the diagnostic techniques used to study the individual processes will be indicated in the present work. The processes that occur during PLAL are plasma formation and expansion ($t \cong 0.1$–$1\ \mu s$), cavitation bubble formation and spread time ($t \cong 1$–$100\ \mu s$), nanoparticle growth ($t \cong 300\ \mu s$), and shockwave emission ($t \cong 1\ \mu s$) [5, 108, 117]. Only the research techniques with appropriate temporal resolution will be listed in this paper, and their more detailed description can be found in the suggested literature. The interferometry ($\sim 10^{-9}$–10^{-6} s) and optical emission spectroscopy ($\sim 10^{-12}$–10^{-9} s) methods to investigate the plasma generation, expansion and quenching mechanism are applied [118, 119]. The shockwave and the cavitation bubble dynamics are studied by fast shadowgraph ($\sim 10^{-9}$–10^{-6} s), X-ray radiography ($\sim 10^{-6}$ s), photoacoustic ($\sim 10^{-6}$ s), and photoelastic imaging ($\sim 10^{-9}$ s). Optical beam deflection ($\sim 10^{-9}$ s), small angle x-ray scattering (SAXS) ($\sim 10^{-6}$ s), laser light scattering ($\sim 10^{-9}$–10^{-6} s), and x-ray radiography ($\sim 10^{-6}$ s) are used to examine the nanoparticle growth. Using techniques such as high-speed photography ($\sim 10^{-10}$–10^{-6} s) and optical beam deflection ($\sim 10^{-9}$ s), the entire dynamics of PLAL procedure can be researched [118–125].

6. Manufacture of NMNPs through laser ablation process

The fabrication of several NMNPs using the PLAL method is covered in this section. The characteristics of the created NMNPs and the corresponding characterization methods will be discussed in detail. Several authors reported the production of NMNPs as Pt, Ag or Au applying the PLAL method in different liquid media – acetone, ethanol and methanol, aqueous solution of sodium dodecyl sulphate (SDS), solution of TSC (trisodium citrate, non-toxic and biocompatible solution) or ethanol in water with different concentrations [71, 72, 74, 126–136].

6.1 Platinum nanoparticles (Pt-NPs)

The Pt-NPs were synthesized in either DDW or different ethanol concentration using an excimer laser, KrF. First, the synthesis of Pt-NPs in DDW will be discussed. To obtain these NPs, the technological parameters were optimized. Three technological parameters were kept constant, and another one was varied to adjust the fabrication of Pt-NPs. The laser fluence was fixed at 2.3 J cm^{-2}, the time ablation was only 15 min, the initial volume of DDW was 6.7 mL every time and the repetition rate was varied from 10 Hz up to 40 Hz. In **Figure 5** are represented the optical properties of the as-created Pt-NPs acquired in the wavelength interval from 200 nm to 800 nm by using the already mentioned technological parameters during the PLAL process.

As can be seen from **Figure 5**, the optical spectrum of the Pt-NPs obtained with 10 Hz RR has the highest transmission (T) value, while the lowest T values were achieved for the Pt-NPs created with 40 Hz RR. In our opinion, these differences between the spectra under consideration it can be due to the difference in the mass concentration of the Pt-NPs in the final solution. Two bands with altered profiles relative to those in the nearby areas may be distinguished in all spectra with reproducible features. This is because of the transmission's lower levels (enhanced extinction values). A distinct local minimum at

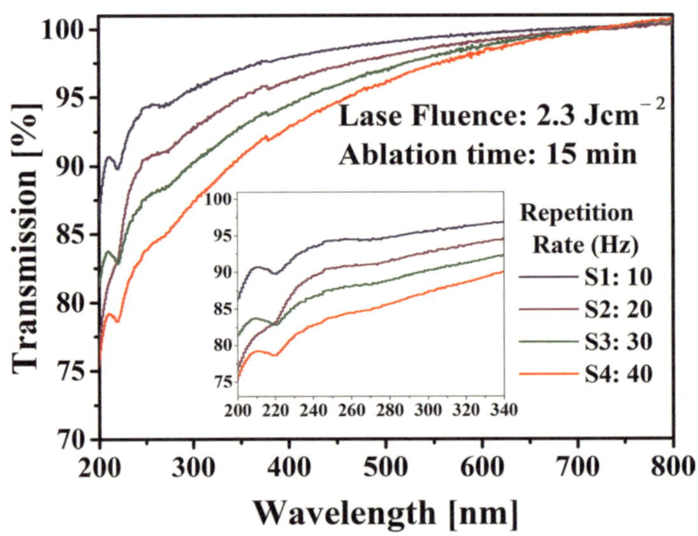

Figure 5.
UV/-Vis measurements of Pt-NPs with 2.3 J cm^{-2}, 15 min ablation time and varied RR: 10 Hz—blue curve, 20 Hz—purple curve, 30 Hz—green curve, and 40 Hz—red curve.

about 220 nm clearly defines the first band which is located in the 200–230 nm range. This band's existence is most likely caused by interband transitions [137–141]. Conversely, a very weakly expressed band with reduced transmission and ambiguous borders can be seen in the wavelength range of 230–330 nm. The local minimum of this band is situated at roughly 270 nm and exhibits relatively modest local minima in all spectra under consideration.

Furthermore, the obtained Pt-NPs were investigated through HR-STEM technique to achieve their morphological properties. **Figure 6** shows the morphological features of the as-prepared Pt-NPs using the above-mentioned technological properties. In addition, the ZC-phase contrast images, **Figure 6a**, and transmission electron (TE) micrographs, **Figure 6b**, which are co-localized images, were simultaneously acquired for each sample at the same magnification (x250K) and position on the sample as a result of a unique characteristic of the HR-STEM used for these measurements.

The below images showed that the Pt-NPs are mostly spherical, and, especially, agglomerated small Pt-NPs exist on the Cu grid. The Pt-NPs obtained at 10 Hz RR have the lowest MS of 2.2 nm, as shown in **Figure 6(Ic)**. When RR is increased up to 20 Hz the MS value rises to 2.8 nm, as is shown in **Figure 2(IIc)**. Further, raising the

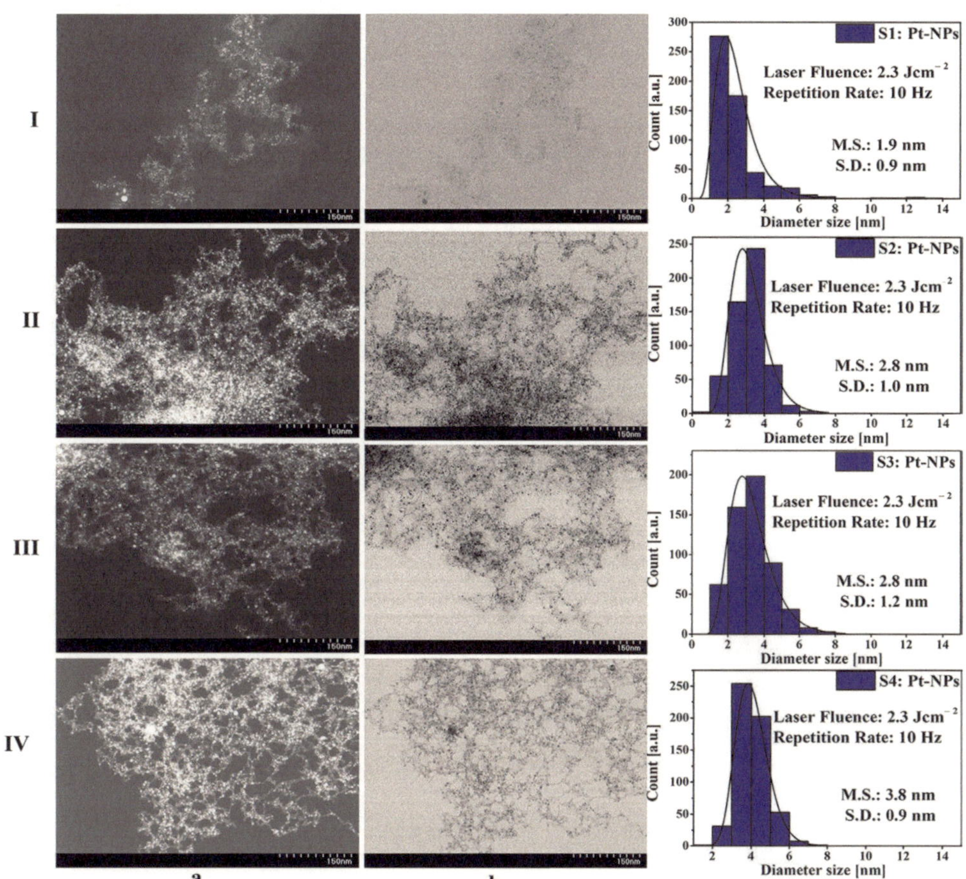

Figure 6.
Co-localized STEM images of obtained Pt-NPs using 2.3 J cm^{-2}: (a) ZC micrographs, (b) TE micrographs and (c) histograms of laser RR: (I) 10 Hz, (II) 20 Hz, (III) 30 Hz, and (IV) 40 Hz.

laser RR to 30 Hz, the MS value remains similar, 2.8 nm, as in the case when 20 Hz RR was used, see **Figure 6(IIIc)**. Nevertheless, the Pt-NPs produced with 40 Hz RR have the largest MS values of 3.8 nm. Regarding the SD values, a decrease ranging from ±1.2 nm (10 Hz RR) to ±0.9 nm (40 Hz RR) can be observed when laser RR is enhanced.

A few research groups produced Pt-NPs in pure water via PLAL method, and they showed that using the lowest values of laser fluence and RR, the smallest mean size value can be obtained. For instance, as mentioned above, Moniri *et al.* obtained the smallest MS value for Pt-NPs using the lowest laser parameters of the Nd:YAG laser [71]. Similarly, Cueto *et al.* produced Pt-NPs in pure water with the fourth harmonic (λ = 266 nm) at RR of 10 Hz and laser fluence of 1.5 J cm^{-2}. They observed a bimodal distribution of Pt-NPs with two MS values of 1–4 nm and 6–8 nm [137]. Nichols *et al.* fabricated Pt-NPs in pure water, by the same laser type with the third harmonic (λ = 355 nm) at laser fluences of 1–110 J cm^{-2}, and the determined diameter size was between 1 and 30 nm [130].

The HR-STEM image of the separated Pt-NPs shows their spherical shape, crystallinity and fully metallic nature as can be seen in **Figure 7a**. This image was used in order to determine the corresponding interplanar profile of d-spacing value measurements for Pt material, see **Figure 7b**. Afterwards, through the white squares created in the transmitted electron micrograph, the interplanar distance of 0.22 nm for the crystalline plane of Pt (111) was established. The achieved value is nearly close to the ideal platinum material with FCC nanostructure value of 0.23 nm [142–144].

In addition, direct analysis in real time-mass spectrometry (DART-MS) was applied to confirm the presence of Pt material in the final solution without supplementary sample preparation of the colloid as a direct investigation.

Figure 8 shows the DART-MS spectrum in which a visible line matching to the Pt-NPs atomic mass can be seen at 195.33 u. where the real atomic mass for Pt is 195.09 u.

In the following part, the synthesis of Pt-NPs in different ethanol concentrations and the manufacturing parameters such as laser fluence (2.3 J cm^{-2}), laser RR (10 Hz), time ablation (10 min), will be discussed. The initial volume of solution (6.7 mL) was fixed while the aqueous ethanol solution was varied: 10, 20, 40, and 60% ethanol concentration. An UV/Vis spectrometer in the wavelength intervals from 200 nm up to 800 nm to obtain the optical spectra of the fabricated Pt-NPs was used. **Figure 9** show the T spectra of the produced Pt-NPs in different ethanol concentrations.

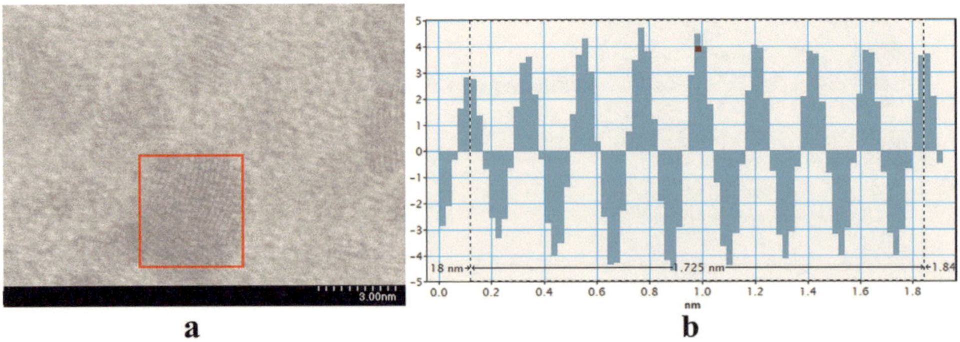

Figure 7.
(a) High resolution STEM micrograph of created Pt-NPs with 2.3 J cm^{-2}, 10 Hz RR and (b) corresponding profile.

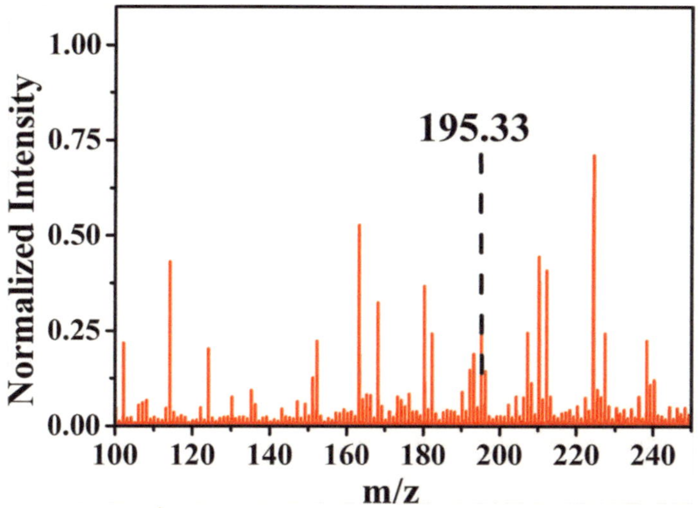

Figure 8.
DART-MS spectrum of Pt-NPs.

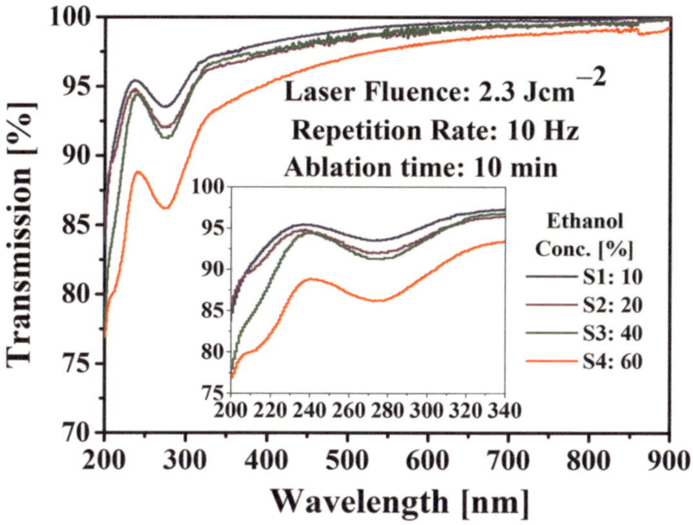

Figure 9.
UV/Vis transmission measurements of Pt-NPs with 2.3 J cm^{-2} laser fluence, 10 Hz RR in different ethanol ratios: 10%—blue curve, 20%—purple curve, 40%—green curve, and 60%—red curve.

As can be seen in **Figure 9**, the Pt-NPs colloids obtained in 10% ethanol concentration have the highest T values than the other Pt-NPs ablated in 20, 40, and 60% aqueous ethanol solution. The SPR in the Pt-NPs is attributed to the optical T band between 230 and 340 nm. As ethanol concentration rises, the minimum position oscillates and the T values of the corresponding spectra decrease. In our opinion, two aspects affect the magnitude of transmission values.

Furthermore, the Pt-NPs' morphological characterization by HR-STEM are shown in **Figure 10** from I to V (a, b), and in **Figure 10c**, the corresponding size distributions are represented.

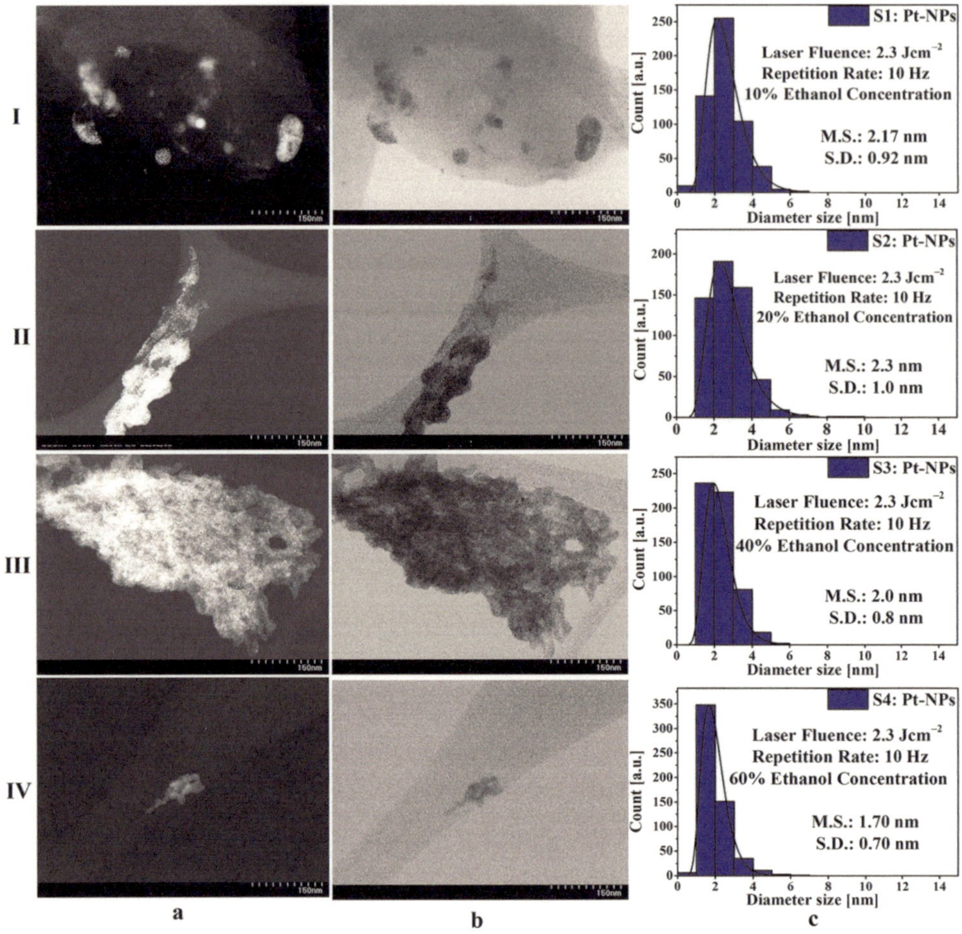

Figure 10.
HR-STEM images of the created Pt-NPs using constant technological parameters but varying ethanol concentration: (I) 10%, (II) 20%, (III) 40%, and (IV) 60%.

All images as can be seen in **Figure 10(I–III (a, b))** show a significant degree of agglomeration in the colloids that are being studied. From S1 to S3 samples, the distinct size of the aggregates gradually rises, the largest aggregate being seen at S3. The specific size of the aggregates rapidly decreases when the ethanol content is increased further, the S4 sample. As can be observed in **Figure 10**(I–IVc), when ethanol concentration is increased from 10% to 60%, the MS values of the Pt-NPs decrease ranging from 2.17 nm to 1.7 nm accompanied by a parallel decrease in SD from 0.92 nm up to 0.7 nm, respectively. Therefore, the smallest MS and the narrowest SD values of 1.7 nm $\pm$ 0.7 nm are established for the Pt-NPs ablated in 60% ethanol concentration with 2.3 J cm^{-2} and 10 Hz RR.

A few scientific groups used either pure ethanol or aqueous ethanol solution as a liquid medium to achieve Pt-NPs with the smallest mean size values [71, 74, 127].

In addition, the X-ray diffraction technique was applied to investigate the structure of the platinum material, and the corresponding diffractogram can be seen in **Figure 11**. The Pt with indexed crystalline planes of (111), (200) and (311), are represented through the peaks at $2\theta = 39.9$, 46.4 and 81.7°, respectively, confirming

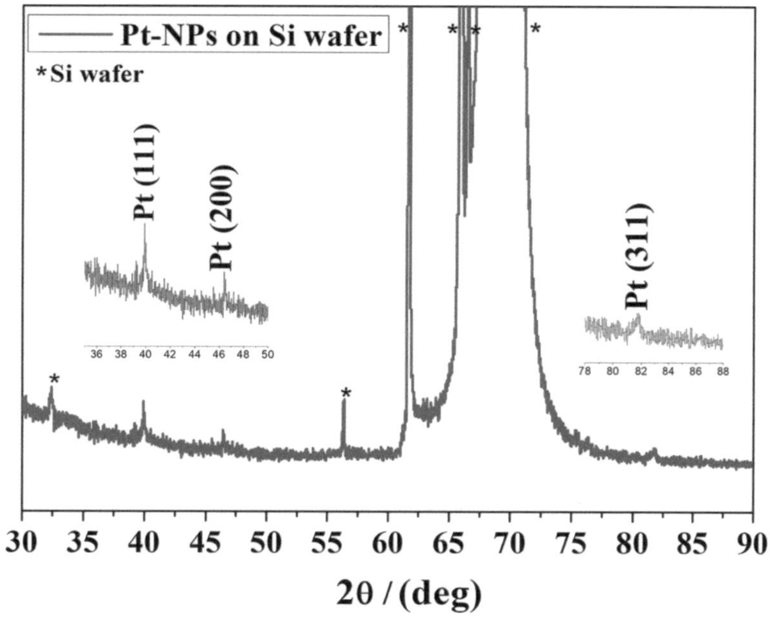

Figure 11.
XRD analysis on fabricated Pt-NPs colloid.

the fabrication of Pt-NPs with nanostructure of FCC. The interplanar distance (d) for the Pt (111), (200) and (311) planes, respectively, were found to be 0.25 nm, 0.19 nm and 0.12 nm [145, 146].

6.2 Gold nanoparticles (Au-NPs)

The Au-NPs were ablated in DDW through of a third harmonic Nd:YAG laser, $\lambda = 355$ nm. The technological parameters such as laser RR (10 Hz), laser ablation duration (20 min) and initial volume of the liquid environment (3.7 mL) were constant, while the laser fluence were varied from 0.9 J cm^{-2} up to 14.5 J cm^{-2}. The laser fluence (the laser spot) was varied in order to obtain different nanostructures from the Au target.

The optical transmission spectra to assess the optical properties of the fabricated nanostructure, as can be seen in **Figure 12**, were acquired in the wavelength interval from 300 nm to 900 nm.

As shown in **Figure 12**, the Au-NPs obtained with the lowest laser fluence has the highest transmission values, while Au colloids fabricated with the highest laser fluence have a significant decrease in transmission values. In our opinion, this effect can happen when the high-density energy of the electron beam is absorbed on the target surface and results in a high quantity of ablated material. Increasing the laser fluence from 0.9 J cm^{-2} to 14.5 J cm^{-2} a red shift of the SPR band maximum from 511 nm to 528 nm can be seen.

Furthermore, the morphological properties of the synthesized nanostructures which are presented in **Figure 13** were recorded by using the Transmission Electron Microscope (TEM) technique.

In **Figure 13(a–d)** different morphologies obtained during PLAL process using different laser fluence values are shown. By using the lowest laser fluence of

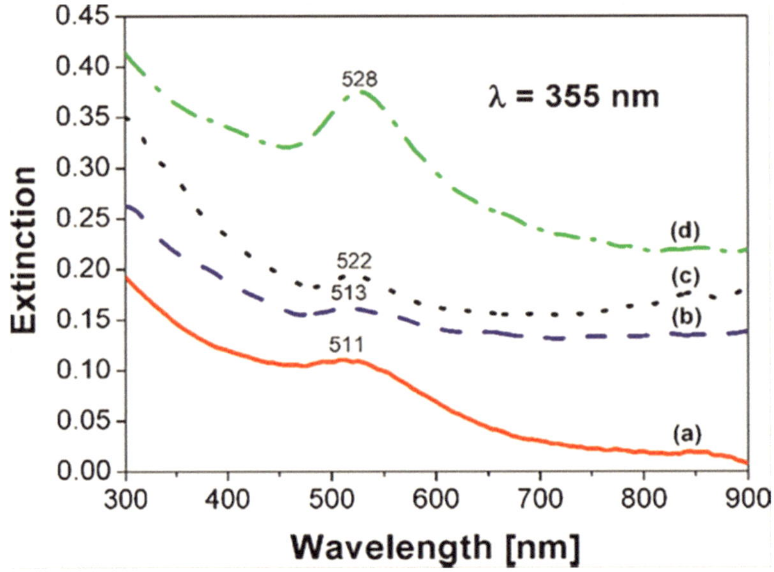

Figure 12.
The UV/Vis extinction spectra on the created nanostructure: (a) 0.9 J cm^{-2}, 2.8 J cm^{-2}, 9.6 J cm^{-2}, and 14.5 J cm^{-2}. Image taken with the permission from [147] and license number 5426541006535.

0.9 J cm^{-2}, **Figure 13a**, Au-NPs of different sizes but having spherical shape with MS value of 4.35 nm can be obtained while increasing the laser fluence value at 2.8 J cm^{-2}, see **Figure 13b**, these Au-NPs begin to be interconnected to each other forming NWs or"nanochains"with a width value of nearly 7 nm. A further increase in the laser fluence at 9.6 J cm^{-2}, as can be observed in **Figure 13c**, the morphological properties of the NWs remain the same as in the previous case, and when the highest laser fluence of 14.5 J cm^{-2} is finally used the NWs begin to disintegrate into Au-NPs with the MS value of 9.3 nm as is shown in **Figure 13d**. In conclusion, using lower and higher laser fluence values, the Au-NPs can be produced with a parallel increase in the MS and SD values from 4.35 $\pm$ 1.6 nm to 9.6 $\pm$ 3.7 nm.

Several results are listed in the following text regarding the fabrication Au-NPs in DDW. Nikolov *et al.* utilized Nd:YAG laser and achieved the best result with the second harmonic wavelength of 532 nm, specifically, MS of 7 nm and SD of 2 nm for Au-NPs [148]. Al-Azawi *et al.* obtained with the fundamental wavelength 13.9 nm, mean particle size with a SD of 6.78 nm. After post-ablation illumination with the second harmonic (532 nm) these results were improved to a mean particle size of 9.50 nm and a SD of 3.12 nm [149]. Jamaludin *et al.* fabricated Au-NPs by the same laser type with the RR of 1 Hz using the fundamental and second harmonic wavelengths i.e., 1064 nm and 532 nm, respectively. They employed 4 different laser energies for each fundamental and second harmonic wavelengths for 30 minutes at 3 Hz pulsed RR. The average MS of the produced NPs is in the range 60–100 nm [150].

6.3 Silver nanoparticles (Ag-NPs)

Another interesting study was done by using PLAL process to ablate and to irradiate the Ag-NPs. The Ag-NPs were ablated through Nd:YAG laser using the fundamental wavelength (λ = 1064 nm) with 10 Hz RR, 12.2 J cm^{-2} laser fluence and

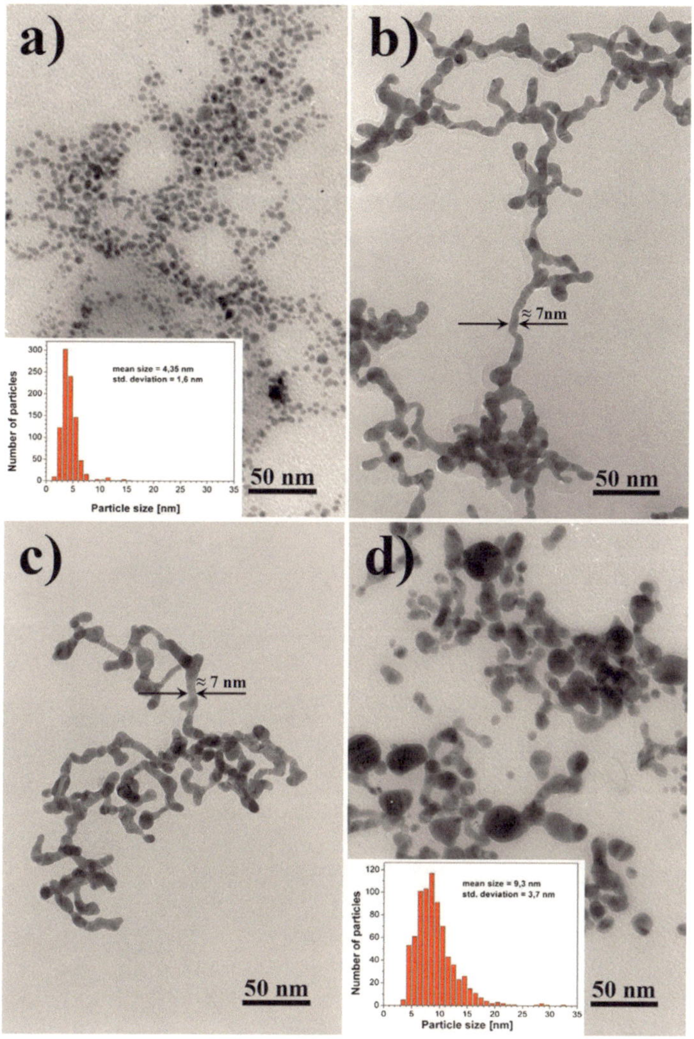

Figure 13.
TEM images of the obtained nanostructures and their corresponding size distributions: (a) 0.9 J cm^{-2}, 2.8 J cm^{-2}, 9.6 J cm^{-2}, and 14.5 J cm^{-2}. Image was taken with permission from [147] and license number 5426541006535.

15 min laser ablation duration. The third harmonic (λ = 355 nm) was used to irradiate the as-prepared colloid to minimize the size distribution. In other words, the laser beam became unfocused using the PLAL setup without a focusing lens. In this case, different values of irradiation time (5–25 min) and pulse laser energy (6 mJ, 17 mJ and 12.5 mJ) were used to optimize the PLAL process to obtain the desired narrowest size distribution. The extinction spectra of the as-prepared and irradiated colloid acquired in the range between 300 and 800 nm are illustrated in **Figure 14a–c**.

As can be seen from **Figure 14**, the transmission values acquired for the unirradiated Ag-NPs in all three cases are lower in the entire wavelength range while their corresponding size distribution is wider but still pronounced. Their size distributions are changed when the Ag-NPs colloids were irradiated at the

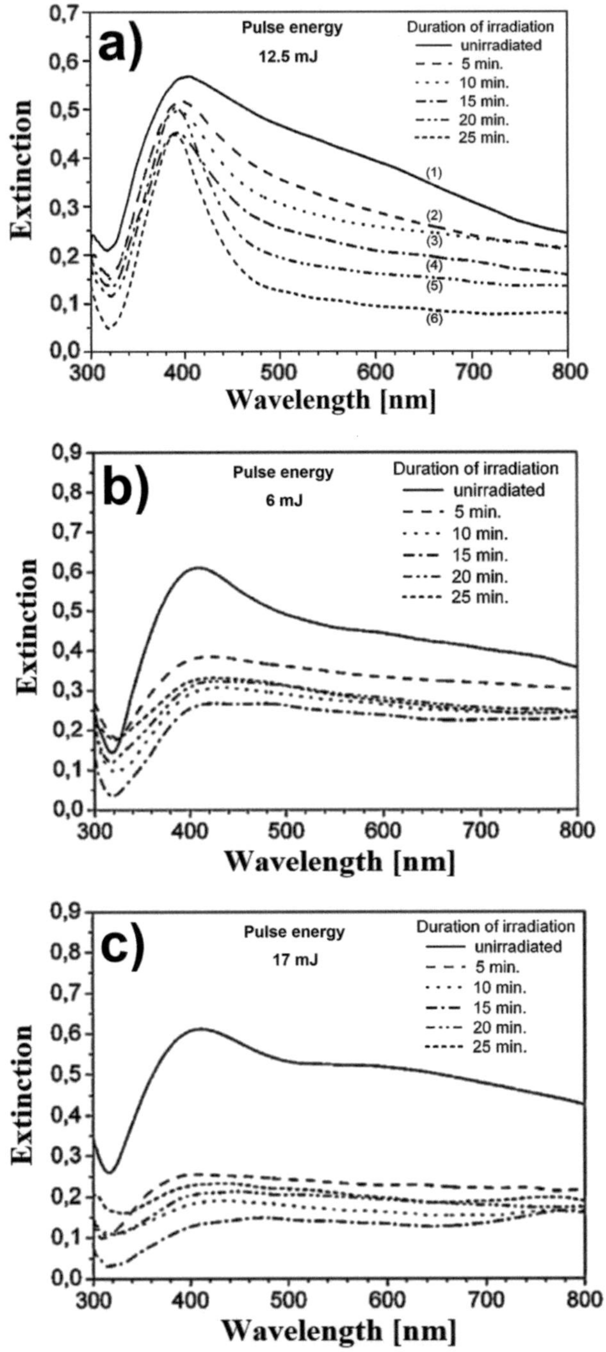

Figure 14.
UV/Vis extinction spectra of the Ag-NPs fabricated with third harmonic, 355 nm, using three different pulse laser energy: (a) 12.5 mJ, (b) 6 mJ and (c) 17 mJ. Image was taken with permission from [151] and license number 5426540770582.

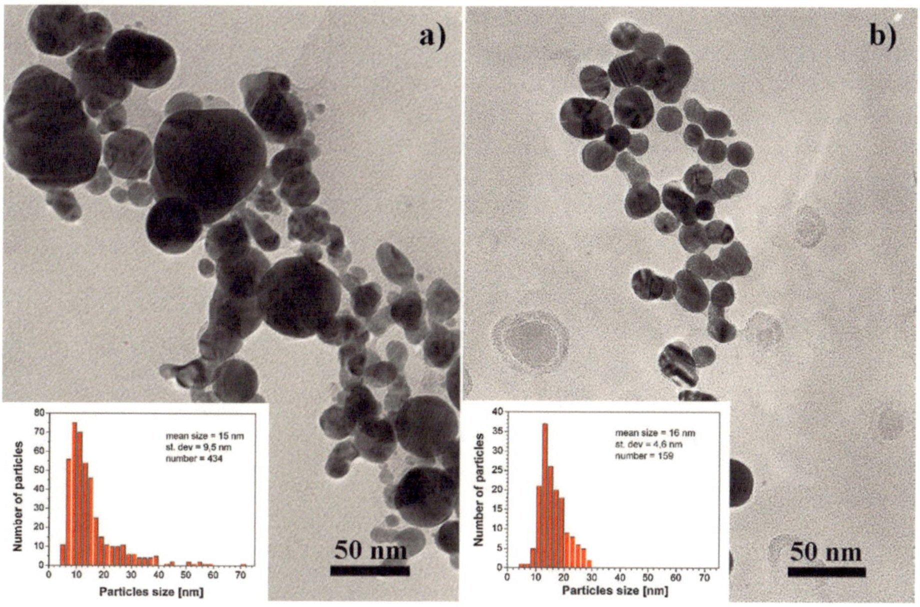

Figure 15.
TEM micrographs of fabricated Ag-NPs: (a) unirradiated and (b) irradiated colloids for 25 min performed using third harmonic and 12.5 mJ laser energy. Image taken with permission from [151] and license number 5426540770582.

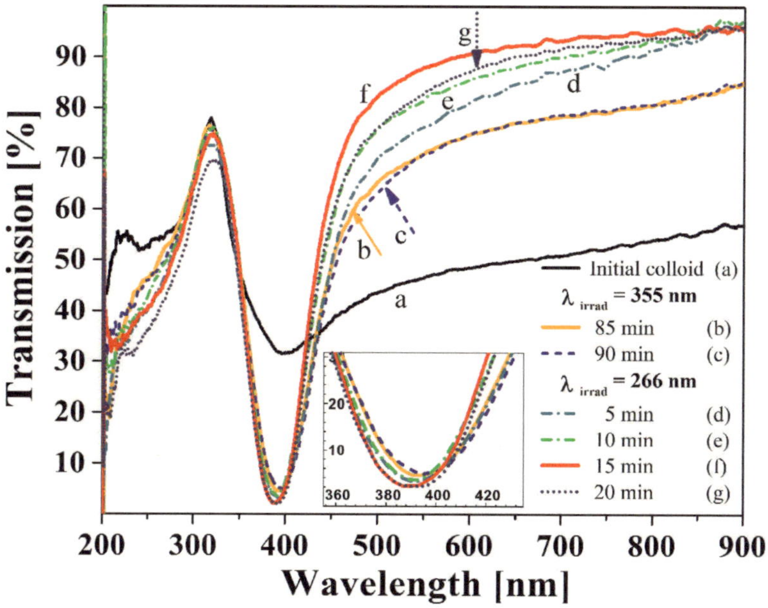

Figure 16.
UV/Vis transmission spectra of Ag-NPs: (a) initial solution immediately after PLAL and laser irradiation with third harmonic after: (b) 85 min, (c) 90 min, and by 266 nm for: (d) 5 min, (e) 10 min, (f) 15 min, (g) 20 min. Image taken with permission from [154] and license number 5426541168853.

same laser pulse energies but using different irradiation times. For instance, in **Figure 14b** and **c**, the NPs size distributions become wider but still noticeable when 6 mJ laser energy is used, while the highest laser energy is utilized the wider and less pronounced size distribution can be obtained. The best result was obtained when the middle value of 12.5 mJ laser energy was used. As can be observed in **Figure 14a**, with increasing the irradiation time the size distribution becomes narrower and more pronounced.

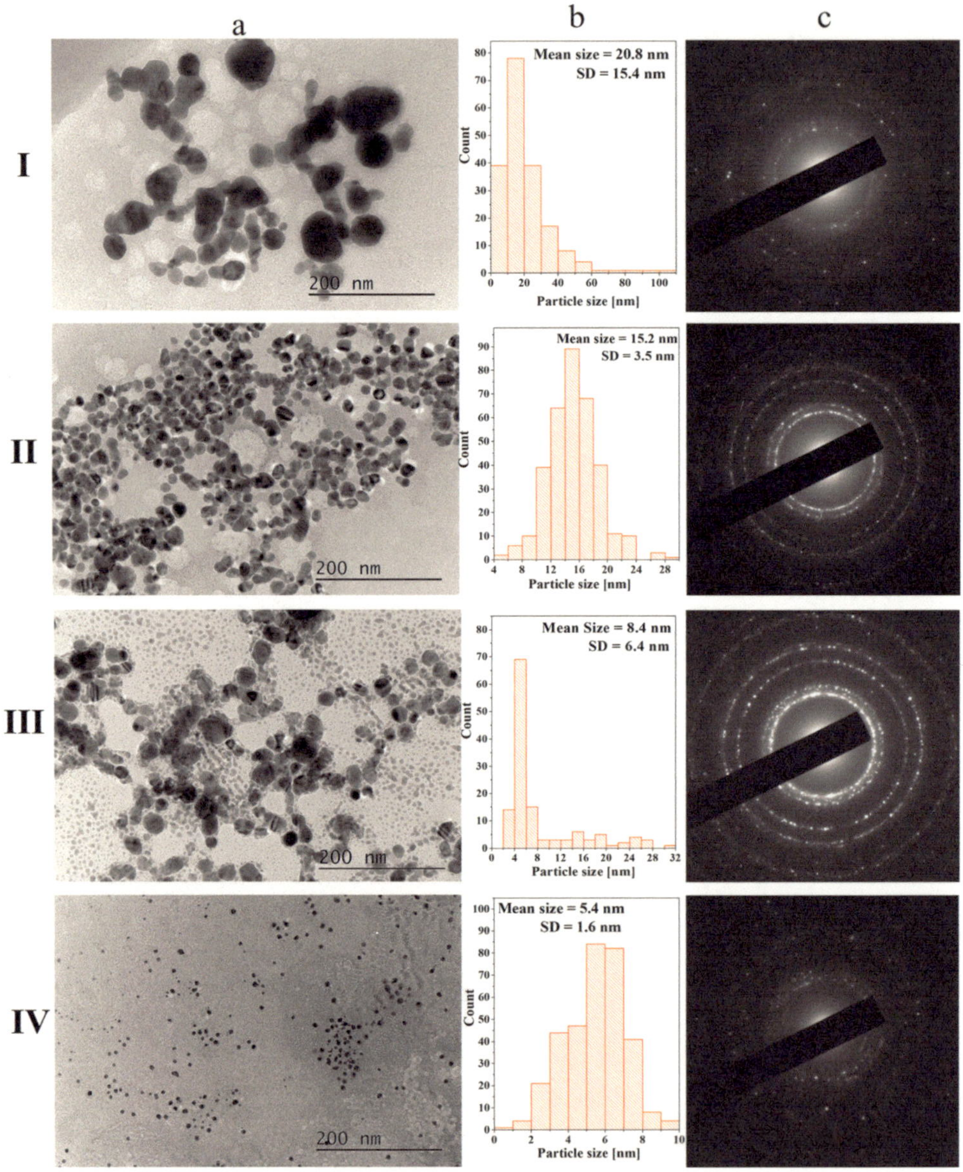

Figure 17.
(a) TEM electron images, (b) histogram graphs, c) SAED patterns of Ag-NPs created after the first phase (I)—immediately after preparation, the second phase (II) with third harmonics—after 85 min, third phase (III) with fourth harmonics (III)—after 40 min, and (IV)—applied third harmonics (after 85 min) and fourth harmonics (after 15 min), respectively. Image was taken with permission from [154] and license number 54265411.

Additionally, the morphological properties were studied through TEM method and are presented in **Figure 15a** and **b**.

Taking a look at the TEM micrographs from **Figure 15**, the unirradiated Ag-NPs are mainly spherical and agglomerated with MS value of 15 nm while those irradiated for 25 min are less agglomerated but the shape differs having the average size of 16 nm. Nevertheless, the TEM results regarding the SD values, ±9.5 nm and ±4.6 nm, respectively, are in good agreement with the UV/Vis results.

The application of PLAL to the synthesis of Ag-NPs is illustrated by the results of several research groups listed below. Tsuji *et al.* investigated the influence of the laser wavelength on particle size [152]. They fabricated colloidal solutions of Ag-NPs in water by ablation with 1064, 532 or 355 nm laser wavelength. The NPs mean diameter became smaller from 29 to 12 nm with decreasing laser wavelength under laser irradiation at the high fluence of 36 J cm^{-2}. Pyatenko *et al.* utilized the second harmonic wavelength (532 nm) [153]. They investigated the influence of the beam spot size, the laser power, and ablation time. Using high laser power and small spot sizes (high laser fluence values), they succeeded in producing very small spherical Ag-NPs with a mean size of 2–5 nm.

A post-ablation treatment is illustrated in **Figure 16**, which aims to maximum reduce the NP's MS and, in parallel, the maximum narrowing of the size distribution. Sequential irradiation of Ag-NPs solution with 355 nm and 266 nm, respectively, was used. As a result of the irradiation and absorption of the light energy, photo-fragmentation takes place. The peculiarity is that the energy at both wavelengths is absorbed by two different mechanisms—plasmonic at 355 nm and interband at 266 nm. The photo-fragmentation process is also different for the two types of absorption. In interband absorption, fragmentation leads to a more significant decrease in the MS of Ag-NPs with a relatively wider size distribution.
In plasmonic absorption, the MS is less affected, but a stronger narrowing of the size distribution occurs. The synergistic effect when applying the considered procedure is expressed in achieving a strong reduction of the average size combined with a strong narrowing of the size distribution. This effect is illustrated by changing

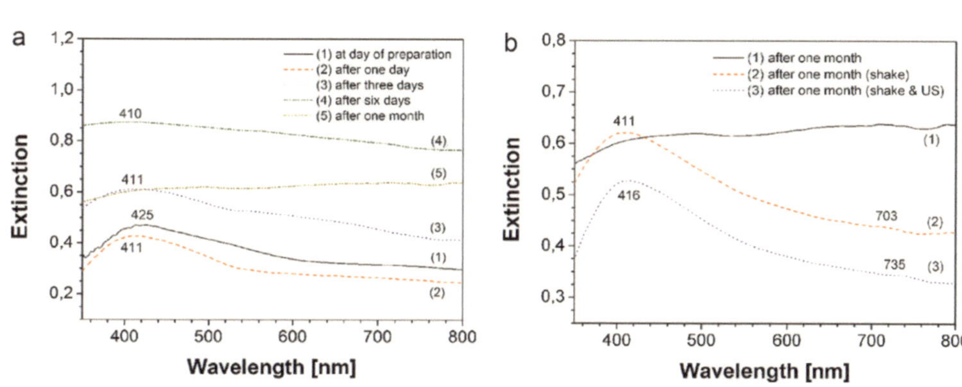

Figure 18.
UV/Vis extinction measurements of Ag-NPs: (a) at different time after preparation ((1) on the preparation day; (2) after 24 h; (3) after 72 h; (4) after 144 h; (5) after 720 h); (b) on the 30th day ((1) initial solution; (2) after manual agitating; (3) after manual agitating followed by a ultrasonic treatment. Image was taken with permission from [155] and license number 5426540458906.

the corresponding optical transmission spectra in **Figure 16** and the TEM images in **Figure 17(I–III)**.

The aging process is expressed in the formation of aggregates and their deposition at the bottom of the cuvette in which they are stored. The change in the optical properties of the aqueous colloid of Ag-NPs within 30 days after the preparation process was followed by the respective optical transmission spectra and is illustrated in **Figure 18a** while in the morphology—by the corresponding TEM images is shown in **Figure 19(I–III)**. The restoration of the colloid's properties, by manual shaking and using ultrasonic treatment, is demonstrated by the optical transmission spectrum, see **Figure 18a** and its TEM image in **Figure 19d**.

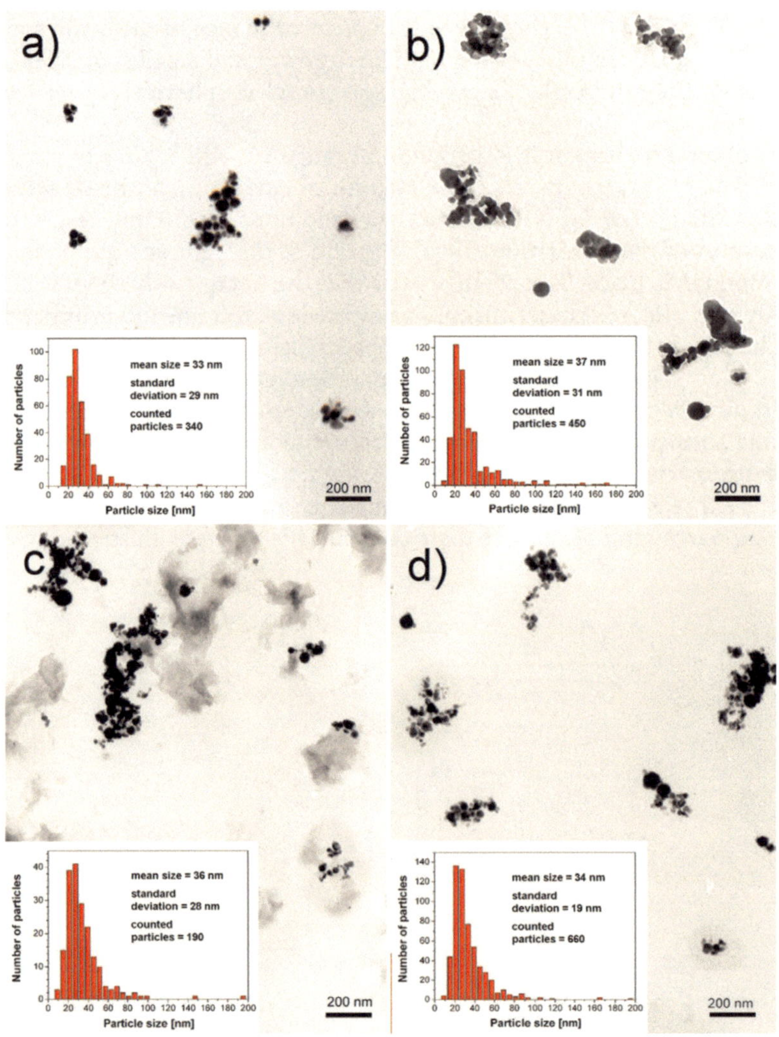

Figure 19.
TEM micrographs of Ag-NPs: (a) on the preparation day; (b) after 720 h; (c) on the 30th day from the upper surface; (d) on the 30th day after manual agitating. Image was taken with permission from [125] and license number 5426540458906.

7. Conclusions

Nanostructures of noble metals like Ag, Au, Pt, Pd, due to their unique antimicrobial, anti-allergic, anti-inflammatory and catalytic properties, find applications in a variety of areas – from nanoelectronics, minimization of environmental pollution, in the industry to biomedicine with an emphasis on cancer treatment. The PLAL method is a versatile top-down physical method to produce them. The synthesized nanoparticles are chemically pure, not containing potentially toxic impurities for the human body, and are therefore biocompatible. Their optical and morphological properties can be largely controlled with an appropriate choice of process parameters such as laser performance and different fluid media. By suitable wavelength, fluence, repetition rate, laser pulse length, type of liquid, and ablation time, nanoparticles on the order of 2–3 nm with a narrow size distribution of the same order can be created. Double-distilled or deionized water is most often used as a liquid medium, but other liquids such as ethanol and its aqueous solution with different concentrations are also used. Controlling the ongoing processes allows more precise handling of the nanostructure's properties. Post-ablation irradiation with the optimal wavelength, fluence and duration of the resulting colloid essentially reduce the average size and narrows the size distribution due to photo-fragmentation. Sequential irradiation of the colloid with different wavelengths enables maximum reduction of the mean size and narrowing of the nanoparticles' size distribution. Such a result is possible using two different absorption mechanisms, plasmonic and interband, which affect the photo-fragmentation process differently. The aging process of silver and gold colloids was studied for 1 month after their creation. Through manual shaking and ultrasonic treatment, the formed aggregates can be destroyed, and the properties of the as-prepared colloids can be restored.

Author contributions

O.A.L., A.S.N. and M.E.—conceptualization; O.A.L., A.S.N., C.C.M. and M.E.—methodology; O.A.L. and A.S.N.—analysis; O.A.L. A.S.N. and C.C.M.—measurements; M.E.—resources; O.A.L., A.S.N. and C.C.M.—data curation; O.A.L. and A.S.N.—writing original draft; O.A.L., A.S.N. and M.E.—writing review and editing; M.E.—supervision; M.E.—project administration; M.E.—funding acquisition.

Funding statement

This research was funded by ECSEL JU under the following grant agreements: No. 826422 (PIn3S), No. 876124 (BEYOND5) and No. 875999 (IT2). The JU receives support from the European Union's Horizon 2020 research and innovation program and Italy, Switzerland, Germany, Belgium, Sweden, Austria, Romania, Slovakia, France, Poland, Spain, Ireland, Israel, Portugal, Greece, Netherlands, Hungary, United Kingdom. This work is financially supported by the Romanian Ministry of Research, Innovation and Digitalization, under the following ECSEL-H2020 Projects: PIn3S-Contract no. 10/1.1.3H/03.04.2020, POC-SMIS code 135127, BEYOND5-Contract no. 12/1.1.3/31.07.2020, POC-SMIS code 136877 and IT2-Contract. no. 11/1.1.3H/06.07.2020, POC-SMIS code 136697.

Conflict of interest

The authors declare no conflict of interest.

Author details

Oana Andreea Lazar[1], Anastas Savov Nikolov[1,2], Călin Constantin Moise[1,3] and Marius Enachescu[1,4]*

1 Center for Surface Science and Nanotechnology, University Politehnica of Bucharest, Bucharest, Romania

2 Institute of Electronics, Bulgarian Academy of Sciences, Sofia, Bulgaria

3 S.C. NanoPRO START MC S.R.L, Pitesti, Romania

4 Academy of Romanian Scientists, Bucharest, Romania

*Address all correspondence to: marius.enachescu@upb.ro

References

[1] Yang GW. Laser ablation in liquids: Applications in the synthesis of nanocrystals. Progress in Materials Science. 2007;**52**:648-698. DOI: 10.1016/j.pmatsci.2006.10.016

[2] Wazeer A, Das A, Sinha A, Karmakar A. Nanomaterials synthesis via laser ablation in liquid: A review. Journal of the Institution of Engineers (India). 2022:1-14. DOI: 10.1007/s40033-022-00370-w

[3] Maiman TH. Stimulated optical radiation in ruby. In: Essentials of Lasers. Hughes Research Laboratories, A Division of Hughes Aircraft Co., Malibu, California: Elsevier; 1969. pp. 134-136. DOI: 10.1016/b978-0-08-013320-1.50016-6

[4] Zhao L, Liu Z, Chen D, Liu F, Yang Z, Li X, *et al.* Laser synthesis and microfabrication of micro/nanostructured materials toward energy conversion and storage. Nano-Micro Letters. 2021;**13**:1-48. DOI: 10.1007/s40820-020-00577-0

[5] Thanh NTK, Maclean N, Mahiddine S. Mechanisms of nucleation and growth of nanoparticles in solution. Chemical Reviews. 2014;**114**:7610-7630. DOI: 10.1021/cr400544s

[6] Boyd IW. Review Paper Thin Film Growth Laser Deposition. Vol. 22. U.K.: Ceramics Internationul; 1996. pp. 429-434

[7] Mafuné F, Kohno JY, Takeda Y, Kondow T. Formation of gold nanonetworks and small gold nanoparticles by irradiation of intense pulsed laser onto gold nanoparticles. Journal of Physical Chemistry B. 2003; **107**:12589-12596. DOI: 10.1021/jp030173l

[8] Corma A, Garci H. Supported gold nanoparticles as catalysts for organic reactions. Chemical Society Reviews. 2008;**37**:2096-2126. DOI: 10.1039/b707314n

[9] Murphy CJ, Gole AM, Stone JW, Sisco PN, Alkilany AM, Goldsmith EC, *et al.* Gold nanoparticles in biology: Beyond toxicity to cellular imaging. Accounts of Chemical Research. 2008; **41**:1721-1730. DOI: 10.1021/ar800035u

[10] Shape-controlled synthesis and surface plasmonic properties of metallic nanostructures. MRS Bulletin. 2005;**30**: 338-343. DOI: 10.1557/mrs2005.96

[11] Amendola V, Meneghetti M. Laser ablation synthesis in solution and size manipulation of noble metal nanoparticles. Physical Chemistry Chemical Physics. 2009;**11**:3805-3821. DOI: 10.1039/b900654k

[12] Caragheorgheopol A, Chechik V. Mechanistic aspects of ligand exchange in Au nanoparticles. Physical Chemistry Chemical Physics. 2008;**10**:5029-5041. DOI: 10.1039/b805551c

[13] Khan I, Saeed K, Khan I. Nanoparticles: Properties, applications and toxicities. Arabian Journal of Chemistry. 2019;**12**:908-931. DOI: 10.1016/j.arabjc.2017.05.011

[14] Amendola V, Meneghetti M. What controls the composition and the structure of nanomaterials generated by laser ablation in liquid solution? Physical Chemistry Chemical Physics. 2013;**15**:3027-3046. DOI: 10.1039/c2cp42895d.

[15] Zhang D, Li Z, Sugioka K. Laser ablation in liquids for nanomaterial

synthesis: Diversities of targets and liquids. Journal of Physics Photonics. 2021;**3**:1-70. DOI: 10.1088/2515-7647/ac0bfd

[16] Fazio E, Gökce B, de Giacomo A, Meneghetti M, Compagnini G, Tommasini M, *et al.* Nanoparticles engineering by pulsed laser ablation in liquids: Concepts and applications. Nanomaterials. 2020;**10**:1-50. DOI: 10.3390/nano10112317

[17] Zotov R, Meshcheryakov E, Livanova A, Minakova T, Magaev O, Isupova L, *et al.* Influence of the composition, structure, and physical and chemical properties of aluminium-oxide-based sorbents on water adsorption ability. Materials. 2018;**11**:1-10. DOI: 10.3390/ma11010132

[18] Leal GF, Lima S, Graça I, Carrer H, Barrett DH, Teixeira-Neto E, *et al.* Design of Nickel Supported on water-tolerant Nb2O5 catalysts for the hydrotreating of lignin streams obtained from lignin-first biorefining. IScience. 2019;**15**:467-488. DOI: 10.1016/j.isci.2019.05.007

[19] Gómez CD, Rodríguez-Páez JE. The effect of the synthesis conditions on structure and photocatalytic activity of Nb2O5 nanostructures. Processing and Application of Ceramics. 2018;**12**: 218-229. DOI: 10.2298/PAC1803218G

[20] Pletnev PM, Pogrebenkov VM, Vereshchagin VI, Tyul'kin DS. Mullite-corundum materials based on mullite binder resistant to high-temperature deformation. Refractories and Industrial Ceramics. 2018;**58**:618-625. DOI: 10.1007/s11148-018-0156-x

[21] Amendola V, Amans D, Ishikawa Y, Koshizaki N, Scirè S, Compagnini G, *et al.* Room-temperature laser synthesis in liquid of oxide, metal-oxide Core-shells, and doped oxide nanoparticles. Chemistry – A European Journal. 2020; **26**:9206-9242. DOI: 10.1002/chem.202000686

[22] Zhang D, Gökce B. Perspective of laser-prototyping nanoparticle-polymer composites. Applied Surface Science. 2017;**392**:991-1003. DOI: 10.1016/j.apsusc.2016.09.150

[23] Zhang D, Gökce B, Barcikowski S. Laser synthesis and processing of colloids: Fundamentals and applications. Chemical Reviews. 2017;**117**:3990-4103. DOI: 10.1021/acs.chemrev.6b00468

[24] Pligovka A, Poznyak A, Norek M. Optical properties of porous alumina assisted niobia nanostructured films–designing 2-d photonic crystals based on hexagonally arranged nanocolumns. Micromachines (Basel). 2021;**12**:1-22. DOI: 10.3390/mi12060589

[25] Said S, Mikhail S, Riad M. Recent processes for the production of alumina nano-particles. Material Science Energy Technology. 2020;**3**:344-363. DOI: 10.1016/j.mset.2020.02.001

[26] Qazi UY, Javaid R. A review on metal nanostructures: Preparation methods and their potential applications. Advanced Nanoparticles. 2016;**05**:27-43. DOI: 10.4236/anp.2016.51004

[27] Gautam A, Komal P, Gautam P, Sharma A, Kumar N, Jung JP. Recent trends in noble metal nanoparticles for colorimetric chemical sensing and micro-electronic packaging applications. Metals (Basel). 2021;**11**:1-21. DOI: 10.3390/met11020329

[28] Krishnapriya R, Nizamudeen C, Saini B, Mozumder MS, Sharma RK, Mourad AHI. MOF-derived Co2+−doped TiO2 nanoparticles as photoanodes for dye-sensitized solar

cells. Scientific Reports. 2021;**11**:1-12. DOI: 10.1038/s41598-021-95844-4

[29] Zhou K, Zhou X, Liu J, Huang Z. Application of magnetic nanoparticles in petroleum industry: A review. Journal of Petroleum Science Engineering. 2001; **188**. DOI: 10.1016/j.petrol.2020.106943

[30] Aliev R, Gulomov J, Abduvohidov M, Aliev S, Ziyoitdinov Z, Yuldasheva N. Stimulation of photoactive absorption of sunlight in thin layers of silicon structures by metal nanoparticles. Applied Solar Energy (English Translation of Geliotekhnika). 2020;**56**:364-370. DOI: 10.3103/S0003701X20050035

[31] Beck A, Huang X, Artiglia L, Zabilskiy M, Wang X, Rzepka P, *et al.* The dynamics of overlayer formation on catalyst nanoparticles and strong metal-support interaction. Natural Communication. 2020;**11**:1-8. DOI: 10.1038/s41467-020-17070-2

[32] Subhan A, Mourad AHI, Al-Douri Y. Influence of laser process parameters, liquid medium, and external field on the synthesis of colloidal metal nanoparticles using pulsed laser ablation in liquid: A review. Nanomaterials. 2022;**12**:1-32. DOI: 10.3390/nano12132144

[33] Fievet F, Ammar-Merah S, Brayner R, Chau F, Giraud M, Mammeri F, *et al.* The polyol process: A unique method for easy access to metal nanoparticles with tailored sizes, shapes and compositions. Chemical Society Reviews. 2018;**47**:5187-5233. DOI: 10.1039/c7cs00777a

[34] Yan K, Xu F, Wei W, Yang C, Wang D, Shi X. Electrochemical synthesis of chitosan/silver nanoparticles multilayer hydrogel coating with pH-dependent controlled release capability and antibacterial property. Colloids Surface B Biointerfaces. 2021;**202**:1-9. DOI: 10.1016/j.colsurfb.2021.111711

[35] Arya S, Mahajan P, Mahajan S, Khosla A, Datt R, Gupta V, *et al.* Review—Influence of processing parameters to control morphology and optical properties of sol-gel synthesized ZnO nanoparticles. ECS Journal of Solid State Science and Technology. 2021;**10**: 023002. DOI: 10.1149/2162-8777/abe095

[36] Yang Y, Johansson M, Wiorek A, Tarakina NV, Sayed F, Mathieu R, *et al.* Gamma-radiation induced synthesis of freestanding nickel nanoparticles. Dalton Transactions. 2021;**50**:376-383. DOI: 10.1039/d0dt03223a

[37] Chin CDW, Treadwell LJ, Wiley JB. Microwave synthetic routes for shape-controlled catalyst nanoparticles and nanocomposites. Molecules. 2021;**26**:1-16. DOI: 10.3390/molecules26123647

[38] Soni V, Raizada P, Singh P, Cuong HN, *et al.* Sustainable and green trends in using plant extracts for the synthesis of biogenic metal nanoparticles toward environmental and pharmaceutical advances: A review. Environmental Research. 2021;**202**:1-18. DOI: 10.1016/j.envres.2021.111622

[39] Ammar HA, el Aty AAA, el Awdan SA. Extracellular myco-synthesis of nano-silver using the fermentable yeasts Pichia kudriavzeviiHA-NY2 and Saccharomyces uvarumHA-NY3, and their effective biomedical applications. Bioprocess and Biosystems Engineering. 2021;**44**:841-854. DOI: 10.1007/s00449-020-02494-3

[40] Salem SS, Fouda A. Green synthesis of metallic nanoparticles and their prospective biotechnological applications: An overview. Biological Trace Element Research. 2011;**199**:

344-370. DOI: 10.1007/s12011-020-02138-3/Published

[41] Odularu AT. Metal nanoparticles: Thermal decomposition, biomedicinal applications to Cancer treatment, and future perspectives. Bioinorganic Chemistry and Applications. 2018;**2018**: 1-6. DOI: 10.1155/2018/9354708

[42] Patil PP, Phase DM, Kulkarni SA, Ghaisas SV, Kulkarni SK, Kanetkar SM, *et al.* Pulsed-Laser-Induced Reactive Quenching at a Liquid-Solid Interface: Aqueous Oxidation of Iron. Phys. Rev. Lett. U.S.1987;**58**:238

[43] Xiao J, Liu P, Wang CX, Yang GW. External field-assisted laser ablation in liquid: An efficient strategy for nanocrystal synthesis and nanostructure assembly. Progress in Materials Science. 2017;**87**:140-220. DOI: 10.1016/j.pmatsci.2017.02.004

[44] Nikov RG, Nedyalkov NN, Karashanova DB. Laser ablation of Ni in the presence of external magnetic field: Selection of microsized particles. Applied Surface Science. 2020;**518**:1-8. DOI: 10.1016/j.apsusc.2020.146211

[45] Liu P, Liang Y, Lin X, Wang C, Yang G. A general strategy to fabricate simple polyoxometalate nanostructures: Electrochemistry-assisted laser ablation in liquid. ACS Nano. 2011;**5**:4748-4755. DOI: 10.1021/nn2007282

[46] John MG. Metal-Oxide Nanostructures Fabricated from Reactive Laser Metal-Oxide Nanostructures Fabricated from Reactive Laser Ablation in Liquid Ablation in Liquid Part of the Materials Chemistry Commons, and the Physical Chemistry Commons Downloaded from Downloaded from. n.d. https://scholarscompass.vcu.edu/etd/6321

[47] Yang S, Cai W, Zeng H, Xu X. Ultra-fine β-SiC quantum dots fabricated by laser ablation in reactive liquid at room temperature and their violet emission. Journal of Materials Chemistry. 2009;**19**: 7119-7123. DOI: 10.1039/b909800c.

[48] Yan Z, Chrisey DB. Pulsed laser ablation in liquid for micro−/nanostructure generation. Journal of Photochemistry and Photobiology C: Photochemistry Reviews. 2012;**13**: 204-223. DOI: 10.1016/j.jphotochemrev.2012.04.004

[49] Kudryashov SI, Samokhvalov AA, Nastulyavichus AA, Saraeva IN, Mikhailovskii VY, Ionin AA, *et al.* Nanosecond-laser generation of nanoparticles in liquids: From ablation through bubble dynamics to nanoparticle yield. Materials. 2019;**12**:1-15. DOI: 10.3390/ma12040562

[50] Kanitz A, Hoppius JS, Gurevich EL, Ostendorf A. Influence of the liquid on femtosecond laser ablation of iron. In: Phys Procedia. Amsterdam, Netherlands: Elsevier B.V; 2016. pp. 114-122. DOI: 10.1016/j.phpro.2016.08.022

[51] Shih CY, Streubel R, Heberle J, Letzel A, Shugaev MV, Wu C, *et al.* Two mechanisms of nanoparticle generation in picosecond laser ablation in liquids: The origin of the bimodal size distribution. Nanoscale. 2018;**10**: 6900-6910. DOI: 10.1039/c7nr08614h

[52] Sugioka K, Cheng Y. Femtosecond laser three-dimensional micro-and nanofabrication. Applied Physical Review. 2014;**1**:1-35. DOI: 10.1063/1.4904320

[53] Basso L, Gorrini F, Bazzanella N, Cazzanelli M, Dorigoni C, Bifone A, *et al.* The modeling and synthesis of nanodiamonds by laser ablation of

graphite and diamond-like carbon in liquid-confined ambient. Applied Physics A Material Science Process. 2018;**124**:1-7. DOI: 10.1007/s00339-017-1491-3

[54] Zhang N, Wang W, Zhu X, Liu J, Xu K, Huang P, *et al.* Investigation of Ultrashort Pulse Laser Ablation of Solid Targets by Measuring the Ablation-Generated Momentum Using a Torsion Pendulum. Costa Rica. 2011

[55] Rafique M, Rafique MS, Kalsoom U, Afzal A, Butt SH, Usman A. Laser ablation synthesis of silver nanoparticles in water and dependence on laser nature. Optical Quantum Electronics. 2019;**51**:1-11. DOI: 10.1007/s11082-019-1902-0

[56] Hikmantiyah N, Hidayanto E, Khumaeni A. Synthesis of Zn-Fe Nanoparticles Using Pulse Laser Ablation as a Contrast Agent in Magnetic Resonance Imaging. Indonesia: Diponegoro University; 2019. Available from: https://ejournal2.undip.ac.id/index.php/jpa/index

[57] Reich S, Göttlicher J, Ziefuss A, Streubel R, Letzel A, Menzel A, *et al.* In situ speciation and spatial mapping of Zn products during pulsed laser ablation in liquids (PLAL) by combined synchrotron methods. Nanoscale. 2020;**12**:14011-14020. DOI: 10.1039/d0nr01500h

[58] Dowding C. Laser ablation. In: Advances in Laser Materials Processing: Technology, Research and Application. Loughborough University, UK. Elsevier Inc.; 2010. pp. 575-628. DOI: 10.1533/9781845699819.7.575

[59] Walsh BM. Nonlinear mixing of Nd:YAG lasers; harmonic and sum frequency generation. Optical Materials (Amst). 2017;**65**:2-7. DOI: 10.1016/j.optmat.2016.07.007

[60] Preuss S, Demchuk A, Stuke M. Sub-Picosecond UV Laser Ablation of Metals. Germany. 1995

[61] Mafuné F, Kohno JY, Takeda Y, Kondow T. Formation of stable platinum nanoparticles by laser ablation in water. Journal of Physical Chemistry B. 2003;**107**:4218-4223. DOI: 10.1021/jp021580k

[62] Mortazavi SZ, Parvin P, Reyhani A, Golikand AN, Mirershadi S. Effect of laser wavelength at IR (1064 nm) and UV (193 nm) on the structural formation of palladium nanoparticles in deionized water. Journal of Physical Chemistry C. 2011;**115**:5049-5057. DOI: 10.1021/jp1091224

[63] Censabella M, Torrisi V, Boninelli S, Bongiorno C, Grimaldi MG, Ruffino F. Laser ablation synthesis of mono- and bimetallic Pt and Pd nanoparticles and fabrication of Pt-Pd/graphene nanocomposites. Applied Surface Science. 2019;**475**:494-503. DOI: 10.1016/j.apsusc.2019.01.029

[64] Nikolov AS, Nedyalkov NN, Nikov RG, Atanasov PA, Alexandrov MT, Karashanova DB. Investigation of Ag nanoparticles produced by nanosecond pulsed laser ablation in water. Applied Physics A: Materials Science & Processing. 2012;**109**:315-322. DOI: 10.1007/s00339-012-7094-0

[65] Shih CY, Shugaev MV, Wu C, Zhigilei LV. The effect of pulse duration on nanoparticle generation in pulsed laser ablation in liquids: Insights from large-scale atomistic simulations. Physical Chemistry Chemical Physics. 2020;**22**:7077-7099. DOI: 10.1039/d0cp00608d

[66] Kabashin AV, Meunier M. Synthesis of colloidal nanoparticles during

femtosecond laser ablation of gold in water. Journal of Applied Physics. 2003; **94**:7941-7943. DOI: 10.1063/1.1626793

[67] Amendola V, Meneghetti M. Controlled size manipulation of free gold nanoparticles by laser irradiation and their facile bioconjugation. Journal of Materials Chemistry. 2007;**17**:4705-4710. DOI: 10.1039/b709621f

[68] Pyatenko A, Wang H, Koshizaki N, Tsuji T. Mechanism of pulse laser interaction with colloidal nanoparticles. Laser Photon Review. 2013;**7**:596-604. DOI: 10.1002/lpor.201300013

[69] Tsuji T, Yahata T, Yasutomo M, Igawa K, Tsuji M, Ishikawa Y, *et al.* Preparation and investigation of the formation mechanism of submicron-sized spherical particles of gold using laser ablation and laser irradiation in liquids. Physical Chemistry Chemical Physics. 2013;**15**:3099-3107. DOI: 10.1039/c2cp44159d

[70] Xu B, Song RG, Wang C. Effect of laser fluence on siliver nanoparticles colloid. Advanced Materials Research. 2012;**2012**:747-750. DOI: 10.4028/www.scientific.net/AMR.415-417.747

[71] Moniri S, Ghoranneviss M, Hantehzadeh MR, Asadabad MA. Synthesis of platinum nanoparticles by nanosecond laser irradiation of bulk Pt in different polar solvents. Research on Chemical Intermediates. 2017;**43**: 3015-3034. DOI: 10.1007/s11164-016-2809-z

[72] Lazar OA, Moise CC, Nikolov AS, Enache LB, Mihai GV, Enachescu M. The water-based synthesis of platinum nanoparticles using KrF excimer laser ablation. Nanomaterials. 2022;**12**:1-18. DOI: 10.3390/nano12030348

[73] Valverde-Alva MA, García-Fernández T, Esparza-Alegría E, Villagrán-Muniz M, Sánchez-Aké C, Castañeda-Guzmán R, *et al.* Laser ablation efficiency during the production of Ag nanoparticles in ethanol at a low pulse repetition rate (1-10 Hz). Laser Physics Letters. 2016; **13**:1-6. DOI: 10.1088/1612-2011/13/10/106002

[74] Lazar OA, Nikolov AS, Moise CC, Rosoiu S, Prodana M, Enachescu M. Fabrication of Pt nanoparticles by nanosecond pulsed laser ablation in aqueous solution of ethanol using KrF excimer laser. Applied Surface Science. 2023;**609**:1-12. DOI: 10.1016/j.apsusc.2022.155289

[75] Altowyan AS, Mostafa AM, Ahmed HA. Effect of liquid media and laser energy on the preparation of Ag nanoparticles and their nanocomposites with Au nanoparticles via laser ablation for optoelectronic applications. Optik (Stuttg). 2021;**241**:1-11. DOI: 10.1016/j.ijleo.2021.167217

[76] Mendivil MI, Krishnan B, Castillo GA, Shaji S. Synthesis and properties of palladium nanoparticles by pulsed laser ablation in liquid. Applied Surface Science. 2015; **348**:45-53. DOI: 10.1016/j.apsusc.2015.03.075

[77] Russo P, Liang R, Jabari E, Marzbanrad E, Toyserkani E, Zhou YN. Single-step synthesis of graphene quantum dots by femtosecond laser ablation of graphene oxide dispersions. Nanoscale. 2016;**8**:8863-8877. DOI: 10.1039/c6nr01148a

[78] Popov AA, Tselikov G, Dumas N, Berard C, Metwally K, Jones N, *et al.* Laser- synthesized TiN nanoparticles as promising plasmonic alternative for biomedical applications. Scientific Reports. 2019;**9**:1-11. DOI: 10.1038/s41598-018-37519-1

[79] Ishikawa Y, Kawaguchi K, Shimizu Y, Sasaki T, Koshizaki N. Preparation of Fe-Pt alloy particles by pulsed laser ablation in liquid medium. Chemical Physics Letters. 2006;**428**: 426-429. DOI: 10.1016/j.cplett.2006.07.076

[80] Nastulyavichus AA, Kudryashov SI, Smirnov NA, Rudenko AA, Kharin AY, Busleev NI, *et al.* Novel approach of controllable stoichiometric fabrication of alloyed Au/Ag nanoparticles by nanosecond laser ablation of thin bi-layered films in water. Laser Physics Letters. 2019;**16**:1-5. DOI: 10.1088/1612-202X/ab35cf

[81] Scaramuzza S, Badocco D, Pastore P, Coral DF, Fernández van Raap MB, Amendola V. Magnetically assembled SERS substrates composed of Iron–silver nanoparticles obtained by laser ablation in liquid. ChemPhysChem. 2017;**18**: 1026-1034. DOI: 10.1002/cphc.201600651

[82] Wagener P, Jakobi J, Rehbock C, Chakravadhanula VSK, Thede C, Wiedwald U, *et al.* Solvent-surface interactions control the phase structure in laser-generated iron-gold core-shell nanoparticles. Scientific Reports. 2016;**6**: 1-12. DOI: 10.1038/srep23352

[83] Marzun G, Bönnemann H, Lehmann C, Spliethoff B, Weidenthaler C, Barcikowski S. Role of dissolved and molecular oxygen on Cu and PtCu alloy particle structure during laser ablation synthesis in liquids. ChemPhysChem. 2017;**18**:1175-1184. DOI: 10.1002/cphc.201601315

[84] Zhang J, Chen G, Guay D, Chaker M, Ma D. Highly active PtAu alloy nanoparticle catalysts for the reduction of 4-nitrophenol. Nanoscale. 2014;**6**:2125-2130. DOI: 10.1039/c3nr04715f

[85] Pramanik A, Karmakar S, Kumbhakar P, Biswas S, Sarkar R, Kumbhakar P. Synthesis of bilayer graphene nanosheets by pulsed laser ablation in liquid and observation of its tunable nonlinearity. Applied Surface Science. 2020;**499**:1-35. DOI: 10.1016/j.apsusc.2019.143902

[86] Al-Douri Y, Al-Samarai RA, Abdulateef SA, Odeh AA, Badi N, Voon CH. Nanosecond pulsed laser ablation to synthesize GaO colloidal nanoparticles: Optical and structural properties. Optik (Stuttg). 2019;**178**: 337-342. DOI: 10.1016/j.ijleo.2018.09.158

[87] Vála L, Vavruňková V, Jandová V, Křenek T. Laser Ablation of Silicon Monoxide and Titanium Monoxide in Liquid: Formation of Mixed Colloidal Dispersion with Photocatalytic Activity. United Kingdom: Institute of Physics Publishing; 2020. DOI: 10.1088/1742-6596/1527/1/012046

[88] Ibrahim YO, Gondal MA, Alaswad A, Moqbel RA, Hassan M, Cevik E, *et al.* Laser-induced anchoring of WO3 nanoparticles on reduced graphene oxide sheets for photocatalytic water decontamination and energy storage. Ceramics International. 2020; **46**:444-451. DOI: 10.1016/j.ceramint.2019.08.281

[89] Anikin KV *et al.* Formation of ZnSe and CdS quantum dots via laser ablation in liquids. 2002. Available from: www.elsevier.com/locate/cplett

[90] Al-Douri Y. Nanosecond pulsed laser ablation to synthesize ternary alloy colloidal nanoparticles. In: Colloidal Metal Oxide Nanoparticles. Amsterdam, Netherlands: Elsevier; 2020. pp. 25-38. DOI: 10.1016/b978-0-12-813357-6.00003-6

[91] Semaltianos NG, Logothetidis S, Perrie W, Romani S, Potter RJ, Sharp M, *et al.* CdSe nanoparticles synthesized by laser ablation. EPL. 2008;**84**:1-6. DOI: 10.1209/0295-5075/84/47001

[92] Darwish AM, Eisa WH, Shabaka AA, Talaat MH. Investigation of factors affecting the synthesis of nano-cadmium sulfide by pulsed laser ablation in liquid environment. Spectrochimica Acta. Part A, Molecular and Biomolecular Spectroscopy. 2016;**153**:315-320. DOI: 10.1016/j.saa.2015.08.007

[93] Ismail RA. Preparation of colloidal lead sulfide nanoparticles by laser ablation in water for optoelectronic devices applications. High Energy Chemistry. 2015;**49**:58-63. DOI: 10.1134/S0018143915010051

[94] Semaltianos NG, Hendry E, Chang H, Wears ML, Monteil G, Assoul M, *et al.* Ns or fs pulsed laser ablation of a bulk InSb target in liquids for nanoparticles synthesis. Journal of Colloid and Interface Science. 2016;**469**: 57-62. DOI: 10.1016/j.jcis.2016.01.069

[95] Qin WJ, Sun J, Yang J, Du XW. Control of Cu-doping and optical properties of ZnO quantum dots by laser ablation of composite targets. Materials Chemistry and Physics. 2011;**130**: 425-430. DOI: 10.1016/j.matchemphys.2011.07.001

[96] Ruth AA, Young JA. Generation of CdSe and CdTe nanoparticles by laser ablation in liquids. Colloids and Surfaces A: Physicochemical and Engineering Aspects. 2006;**279**:121-127. DOI: 10.1016/j.colsurfa.2005.12.049

[97] Al-Mamun SA, Nakajima R, Ishigaki T. Effect of liquid level and laser power on the formation of spherical alumina nanoparticles by nanosecond laser ablation of alumina target. Thin Solid Films. 2012;**2012**:46-51. DOI: 10.1016/j.tsf.2012.06.011

[98] Barcikowski S, Meńndez-Manjón A, Chichkov B, Brikas M, Račiukaitis G. Generation of nanoparticle colloids by picosecond and femtosecond laser ablations in liquid flow. Applied Physics Letters. 2007;**91**:1-3. DOI: 10.1063/1.2773937

[99] Baladi A, Sarraf Mamoory R. Investigation of different liquid media and ablation times on pulsed laser ablation synthesis of aluminum nanoparticles. Applied Surface Science. 2010;**256**:7559-7564. DOI: 10.1016/j.apsusc.2010.05.103

[100] Jiang Y, Liu P, Liang Y, Li HB, Yang GW. Promoting the yield of nanoparticles from laser ablation in liquid. Applied Physics A: Materials Science & Processing. 2011;**105**:903-907. DOI: 10.1007/s00339-011-6557-z

[101] Cristoforetti G, Pitzalis E, Spiniello R, Ishak R, Giammanco F, Muniz-Miranda M, *et al.* Physico-chemical properties of Pd nanoparticles produced by pulsed laser ablation in different organic solvents. Applied Surface Science. 2012;**258**:3289-3297. DOI: 10.1016/j.apsusc.2011.11.084

[102] Moura CG, Pereira RSF, Andritschky M, Lopes ALB, *et al.* Effects of laser fluence and liquid media on preparation of small Ag nanoparticles by laser ablation in liquid. Optical Laser Technology. 2017;**97**:20-28. DOI: 10.1016/j.optlastec.2017.06.007

[103] Lee SH, Jung HJ, Lee SJ, Theerthagiri J, Kim TH, Choi MY. Selective synthesis of Au and graphitic carbon-encapsulated Au (Au@GC) nanoparticles by pulsed laser ablation in solvents: Catalytic Au and acid-resistant Au@GC nanoparticles. Applied Surface

Science. 2020;**506**. DOI: 10.1016/j.apsusc.2019.145006

[104] Kanitz A, Hoppius JS, Fiebrandt M, Awakowicz P, Esen C, Ostendorf A, *et al.* Impact of liquid environment on femtosecond laser ablation. Applied Physics A Material Science Process. 2017; **123**:1-7. DOI: 10.1007/s00339-017-1280-z

[105] Zeng H, Du XW, Singh SC, Kulinich SA, Yang S, He J, *et al.* Nanomaterials via laser ablation/irradiation in liquid: A review. Advanced Functional Materials. 2012;**22**:1333-1353. DOI: 10.1002/adfm.201102295

[106] Dell'Aglio M, Gaudiuso R, Elrashedy R, de Pascale O, Palazzo G, de Giacomo A. Collinear double pulse laser ablation in water for the production of silver nanoparticles. Physical Chemistry Chemical Physics. 2013;**15**:20868-20875. DOI: 10.1039/c3cp54194k

[107] de Giacomo A, Dell'Aglio M, Santagata A, Gaudiuso R, de Pascale O, Wagener P, *et al.* Cavitation dynamics of laser ablation of bulk and wire-shaped metals in water during nanoparticles production. Physical Chemistry Chemical Physics. 2013;**15**:3083-3092. DOI: 10.1039/c2cp42649h

[108] de Giacomo A, Dell'Aglio M, de Pascale O, Gaudiuso R, Palleschi V, Parigger C, *et al.* Plasma processes and emission spectra in laser induced plasmas: A point of view. Spectrochim Acta Part B At Spectroscopic. 2014;**100**: 180-188. DOI: 10.1016/j.sab.2014.08.013

[109] Casavola A, de Giacomo A, Dell'Aglio M, Taccogna F, Colonna G, de Pascale O, *et al.* Experimental investigation and modelling of double pulse laser induced plasma spectroscopy under water. Spectrochim Acta Part B At Spectroscopic. 2005;**2005**:975-985. DOI: 10.1016/j.sab.2005.05.034

[110] de Giacomo A, de Bonis A, Dell'aglio M, de Pascale O, Gaudiuso R, Orlando S, *et al.* Laser ablation of graphite in water in a range of pressure from 1 to 146 atm using single and double pulse techniques for the production of carbon nanostructures. Journal of Physical Chemistry C. 2011; **115**:5123-5130. DOI: 10.1021/jp109389c

[111] de Giacomo A, Dell'Aglio M, Bruno D, Gaudiuso R, de Pascale O. Experimental and theoretical comparison of single-pulse and double-pulse laser induced breakdown spectroscopy on metallic samples. Spectrochim Acta Part B At Spectrosc. 2008;**63**:805-816. DOI: 10.1016/j.sab.2008.05.002

[112] Ibrahimkutty S, Wagener P, Menzel A, Plech A, Barcikowski S. Nanoparticle formation in a cavitation bubble after pulsed laser ablation in liquid studied with high time resolution small angle x-ray scattering. Applied Physics Letters. 2012;**101**:1-4. DOI: 10.1063/1.4750250

[113] Byun KT, Kwak HY. A model of laser-induced cavitation. Japanese Journal of Applied Physics, Part 1: Regular Papers and Short Notes and Review Papers. 2004;**43**:621-630. DOI: 10.1143/JJAP.43.621

[114] Kotaidis V, Plech A. Cavitation dynamics on the nanoscale. Applied Physics Letters. 2005;**87**:1-3. DOI: 10.1063/1.2132086

[115] Sasaki K, Nakano T, Soliman W, Takada N. Effect of pressurization on the dynamics of a cavitation bubble induced by liquid-phase laser ablation. Applied Physics Express. 2009;**2**: 0465011-0465013. DOI: 10.1143/APEX.2.046501

[116] DR Objectives, Behind the Research. n.d

[117] Tomko J, O'Malley SM, Trout C, Naddeo JJ, Jimenez R, Griepenburg JC, *et al.* Cavitation bubble dynamics and nanoparticle size distributions in laser ablation in liquids. Colloids and Surfaces A: Physicochemical and Engineering Aspects. 2017;**522**:368-372. DOI: 10.1016/j.colsurfa.2017.03.030

[118] Schodel R. Practical realisation of the length by interferometry-general principles and limitations. In: Modern Interferometry for Length Metrology: Exploring Limits and Novel Techniques. United Kingdom: IOP Publishing; 2018. DOI: 10.1088/2053-2563/aadddcch1

[119] Bruggeman P, Verreycken T, González MÁ, Walsh JL, Kong MG, Leys C, *et al.* Optical emission spectroscopy as a diagnostic for plasmas in liquids: Opportunities and pitfalls. Journal of Physics D Applied Physics. 2010;**43**:1-8. DOI: 10.1088/0022-3727/43/12/124005

[120] Ostermayr TM, Kreuzer C, Englbrecht FS, Gebhard J, Hartmann J, Huebl A, *et al.* Laser-driven x-ray and proton micro-source and application to simultaneous single-shot bi-modal radiographic imaging. Nature Communication. 2020;**11**:1-9. DOI: 10.1038/s41467-020-19838-y

[121] Thréard T, de Lima Savi E, Avanesyan S, Chigarev N, Hua Z, Tournat V, *et al.* Photoacoustic 3-D imaging of polycrystalline microstructure improved with transverse acoustic waves. Photoacoustics. 2021;**23**:1-21. DOI: 10.1016/j.pacs.2021.100286

[122] Bothell JK, Machicoane N, Li D, Morgan TB, Aliseda A, Kastengren AL, *et al.* Comparison of X-ray and optical measurements in the near-field of an optically dense coaxial air-assisted atomizer. International Journal of Multiphase Flow. 2020;**125**:1-9. DOI: 10.1016/j.ijmultiphaseflow.2020.103219

[123] Eriksson I. New high-speed photography technique for observation of fluid flow in laser welding. Optical Engineering. 2010;**49**:100503. DOI: 10.1117/1.3502567

[124] Putman CAJ, de Grooth BG, van Hulst NF, Greve J. A detailed analysis of the optical beam deflection technique for use in atomic force microscopy. Journal of Applied Physics. 1992;**72**:6-12. DOI: 10.1063/1.352149

[125] Ishmatov AN, Akhmadeev IR. The low-angle laser light scattering method in the study of pulse liquid atomization. Atmospheric and Oceanic Optics. 2013; **26**:444-448. DOI: 10.1134/S1024856013050084

[126] Calderón-Jiménez B, Johnson ME, Montoro Bustos AR, Murphy KE, Winchester MR, Baudrit JRV. Silver nanoparticles: Technological advances, societal impacts, and metrological challenges. Frontiers in Chemistry. 2017; **5**:1-26. DOI: 10.3389/fchem.2017.00006

[127] Mendivil Palma MI, Krishnan B, Rodriguez GAC, Roy TKD, Avellaneda DA, Shaji S. Synthesis and properties of platinum nanoparticles by pulsed laser ablation in liquid. Journal of Nanomaterials. 2016;**2016**:1-11. DOI: 10.1155/2016/9651637

[128] Menazea AA. Femtosecond laser ablation-assisted synthesis of silver nanoparticles in organic and inorganic liquids medium and their antibacterial efficiency. Radiation Physics and Chemistry. 2020;**168**:1-13. DOI: 10.1016/j.radphyschem.2019.108616

[129] Ganash EA, Altuwirqi RM. Size control of synthesized silver

nanoparticles by simultaneous chemical reduction and laser fragmentation in origanum majorana extract: Antibacterial application. Materials. 2021;**14**:1-11. DOI: 10.3390/ma14092326

[130] Nichols WT, Sasaki T, Koshizaki N. Laser ablation of a platinum target in water. II. Ablation rate and nanoparticle size distributions. Journal of Applied Physics. 2006;**100**:1-6. DOI: 10.1063/1.2390641

[131] Gentile L, Mateos H, Mallardi A, Dell'Aglio M, de Giacomo A, Cioffi N, *et al.* Gold nanoparticles obtained by ns-pulsed laser ablation in liquids (ns-PLAL) are arranged in the form of fractal clusters. Journal of Nanoparticle Research. 2020;**23**:1-12. DOI: 10.1007/s11051-021-05140-5

[132] Isa SZM, Zainon R, Tamal M. State of the art in gold nanoparticle synthesisation via pulsed laser ablation in liquid and its characterisation for molecular imaging: A review. Materials. 2022;**15**:1-20. DOI: 10.3390/ma15030875

[133] Khalaf Ali A. Preparation of Gold Nanoparticles by Pulsed Laser Ablation in NaOH Solution. n.d

[134] Pal S, Tak YK, Song JM. Does the antibacterial activity of silver nanoparticles depend on the shape of the nanoparticle? A study of the gram-negative bacterium Escherichia coli. Applied and Environmental Microbiology. 2007;**73**:1712-1720. DOI: 10.1128/AEM.02218-06

[135] Pulsed Laser Ablation in Liquid for Gold Nanoparticle. n.d

[136] Nguyen TB, Nguyen TD, Nguyen QD, Nguyen TT. Preparation of platinum nanoparticles in liquids by laser ablation method. Advances in Natural Sciences: Nanoscience and Nanotechnology. 2014;**5**. DOI: 10.1088/2043-6262/5/3/035011

[137] Cueto M, Sanz M, Oujja M, Gámez F, Martínez-Haya B, Castillejo M. Platinum nanoparticles prepared by laser ablation in aqueous solutions: Fabrication and application to laser desorption ionization. Journal of Physical Chemistry C. 2011;**115**:22217-22224. DOI: 10.1021/jp207324h

[138] Fernández-Hernández RC, Gleason-Villagran R, Torres-Torres C, Rodríguez-Fernández L, Crespo-Sosa A, Cheang-Wong JC, *et al.* On the physical contributions to the third-order nonlinear optical response in plasmonic nanocomposites. Journal of Optics (United Kingdom). 2012;**14**:1-10. DOI: 10.1088/2040-8978/14/12/125203

[139] Creighton JA, Eadont DG. Ultraviolet-Visible Absorption Spectra of the Colloidal Metallic Elements. United Kingdom. 1991

[140] Bigall NC, Härtling T, Klose M, Simon P, Eng LM, Eychmüller A. Monodisperse platinum nanospheres with adjustable diameters from 10 to 100 nm: Synthesis and distinct optical properties. Nano Letters. 2008;**8**:4588-4592. DOI: 10.1021/nl802901t

[141] Davis-Wheeler Chin C, Akbarian-Tefaghi S, Reconco-Ramirez J, Wiley JB. Rapid microwave synthesis and optical activity of highly crystalline platinum nanocubes. MRS Communication. 2018;**8**:71-78. DOI: 10.1557/mrc.2017.137

[142] Hiramatsu M, Hori M. Preparation of dispersed platinum nanoparticles on a carbon nanostructured surface using supercritical fluid chemical deposition. Materials. 2010;**3**:1559-1572. DOI: 10.3390/ma3031559

[143] Liang S, Xia Y, Zhu S, Zheng S, He Y, Bi J, *et al.* Au and Pt co-loaded g-C 3 N 4 nanosheets for enhanced photocatalytic hydrogen production under visible light irradiation. Applied Surface Science. 2015;**2015**: 304-312. DOI: 10.1016/j.apsusc.2015.08.035

[144] Ma YW, Liu ZR, Wang BL, Zhu L, Yang JP, Li XA. Preparation of graphene-supported Pt-Co nanoparticles and their use in oxygen reduction reactions. Xinxing Tan Cailiao/New Carbon Materials. 2012;**27**:250-257. DOI: 10.1016/S1872-5805(12)60016-X

[145] Kumar MN, Govindh B, Annapurna N. Green synthesis and characterization of platinum nanoparticles using Sapindus mukorossi Gaertn fruit pericarp. Asian Journal of Chemistry. 2017;**29**:2541-2544. DOI: 10.14233/ajchem.2017.20842A

[146] Ji W, Qi W, Tang S, Peng H, Li S. Hydrothermal synthesis of ultrasmall Pt nanoparticles as highly active electrocatalysts for methanol oxidation. Nanomaterials. 2015;**5**:2203-2211. DOI: 10.3390/nano5042203

[147] Nikolov AS, Nedyalkov NN, Nikov RG, Dimitrov IG, Atanasov PA, Maximova K, *et al.* Processing conditions in pulsed laser ablation of gold in liquid for fabrication of nanowire networks. Applied Surface Science. 2014;**2014**: 243-249. DOI: 10.1016/j.apsusc.2014.02.010

[148] Nikolov AS, Nedyalkov NN, Nikov RG, Atanasov PA, Alexandrov MT. Characterization of Ag and Au nanoparticles created by nanosecond pulsed laser ablation in double distilled water. Applied Surface Science. 2011;**2011**:5278-5282. DOI: 10.1016/j.apsusc.2010.10.146

[149] Al-Azawi MA, Bidin N. Gold nanoparticles synthesized by laser ablation in deionized water. Chinese Journal of Physics. 2015; **53**:201-209. DOI: 10.6122/CJP.20150511B

[150] Jamaludin N, Chaudhary KT, Haider Z, Duralim M, Ismail FD, Roslan MS, *et al.* Effect of laser energy and wavelength on average size of gold nanoparticles synthesized by pulsed laser ablation in deionized water. In: J Phys Conf Ser. United Kingdom: IOP Publishing Ltd; 2020. DOI: 10.1088/1742-6596/1484/1/012029

[151] Nikolov AS, Nikov RG, Dimitrov IG, Nedyalkov NN, Atanasov PA, Alexandrov MT, *et al.* Modification of the silver nanoparticles size-distribution by means of laser light irradiation of their water suspensions. Applied Surface Science. 2013; **280**:55-59. DOI: 10.1016/j.apsusc.2013.04.079

[152] Tsuji T, Iryo K, Watanabe N, Tsuji M. Preparation of Silver Nanoparticles by Laser Ablation in Solution: Influence of Laser Wavelength on Particle Size. 2023

[153] Pyatenko A, Shimokawa K, Yamaguchi M, Nishimura O, Suzuki M. Synthesis of silver nanoparticles by laser ablation in pure water. In: Appl Phys A Mater Sci Process. Springer Verlag; 2004. pp. 803-806. DOI: 10.1007/s00339-004-2841-5

[154] Nikolov AS, Stankova NE, Karashanova DB, Nedyalkov NN, Pavlov EL, Koev KT, *et al.* Synergistic effect in a two-phase laser procedure for production of silver nanoparticles colloids applicable in ophthalmology.

Optical Laser Technology. 2021;**138**:1-8. DOI: 10.1016/j.optlastec.2020.106850

[155] Nikov RG, Nikolov AS, Nedyalkov NN, Dimitrov IG, Atanasov PA, Alexandrov MT. Stability of contamination-free gold and silver nanoparticles produced by nanosecond laser ablation of solid targets in water. Applied Surface Science. 2012;**2012**: 9318-9322. DOI: 10.1016/j.apsusc.2011.12.040

Chapter 3

Study of Optical and Structural Properties of ZnS/rGO Nano Composites Prepared by Laser Ablation

Mohammed J. Jader and Wasan H. Jawad

Abstract

Through the physical process of laser ablation, zinc sulfide (ZnS/rGO) nanoparticles were created. From there, it was possible to estimate the crystal structure and crystallite size of X-ray diffraction (XRD) particles (D = 8.35 nm). The visible ultraviolet absorption spectrum has also been used to study optical characteristics. The Neodymium-Yak laser (1064 nm) with the pulsed laser energy of 800 mJ and 600 pulse at a frequency of 6 Hz in room temperature was used to form the Graphen Oxide as a Nano composite after it had been prepared by the hummer method and combined with zinc sulfide. This resulted in the generation of plasma that contributed to the formation of Nanoparticles. To determine the components of the material, several measurements were done on it, including FTIR, SEM, XRD and EDAX for samples. In this chapter, ZnS/rGO nanoparticles were prepared by laser ablation method as well as by hydrothermal chemical method, and optical and structural examinations for both were found. Hence, the laser ablation method proved to be highly efficient compared to the chemical method in terms of knowing the Nano scale sizes and the different energy gaps, which helps in finding multiple future applications for them.

Keywords: nano composite, ZnS nano materials, laser ablation, Zns/rGO, nano particles

1. Introduction

A variety of bottom-up production techniques for nanoparticles have been created for both the liquid (such as sol-gel and chemical reduction) and vapor phases (such as physical/chemical vapor deposition and flame synthesis). Each fabrication technique has benefits and drawbacks. For the purpose of creating different types of nanoparticles with precisely regulated structures on a laboratory scale, liquid phase technologies are efficient and utilised. By using a continuous flow reactor, vapor phase techniques are superior at creating high purity nanoparticles. Solid nanoparticles are produced by the nucleation of supersaturated species that are created by precursor reactions and/or the evaporation of solids in both the liquid and gas bottom-up processes. According to Buzea et al. [1], nanoparticles exhibit some distinctive qualities that are not seen in bulk materials. According to Kim et al. [2], the most important property of nanoparticles is that it greatly depends on the size and size distribution of the particles in order to have specific properties (electrical, optical, magnetic, etc.).

This chapter will present Preparation of graphene oxide (GO) and ZnS disc with using laser ablation method, the results of the optical examinations obtained through the UV-Vis spectrometer and the synthetic through XRD devices (FTIR, SEM, EDX) of the ZnS/rGO nanocomposite prepared by the laser ablation method. Liquid phase pulse (PL-AL) and (hydrothermal) method are discussed.

2. Preparation of graphene oxide (GO)

Graphene oxide was prepared using the Hummer modified method, which is considered one of the most common methods in the production of large quantities of graphene oxide (1 g) of pure and fine graphite powder was added and (120 ml) of concentrated sulfuric acid H_2SO_4, and (13.3 ml) of graphene oxide. H_3PO_4 phosphoric acid, and stir the mixture using a magnetic stirrer. Then (6 g) of $KMnO_4$ potassium permanganate was gradually added to the mixture, and the mixture was stirred at a temperature of (50°C) for (12) hours. Hours, we notice the color change of the mixture to a dark brown color, and after 10 h it changes to a light brown color, and after completing (12) hours, (4 ml) of hydrogen peroxide H_2O_2 was added slowly in an ice bath, we notice the color change of the mixture from the color Light brown to yellow, then the ice bath was removed, the color of the mixture changed indicating the formation of graphene oxide (GO), then the graphene oxide was separated using a centrifuge at 8000 rpm for 15 min, where it was washed for the first time using (30%) of acid HCL (2–3 times), then it was washed with (5%) of acid HCL (2–3 times) also, then distilled water was used repeatedly (8–10 times) until the number becomes The pH of the solution is neutral 6–7), then it was washed twice with ethanol, then washing with (40 ml) of ether and only once, after that The resulting suspension was filtered using filter paper and dried at room temperature in order to obtain graphene oxide (GO) powder.

3. Preparation of the ZnS/rGO nanocomposite

ZnS/rGO was prepared by two methods, physical (pulsed laser ablation) and chemical (hydrothermal).

3.1 Preparation of (GO) solution

Graphene oxide (GO) powder (70 mg), prepared in the aforementioned manner in paragraph and weighed and dissolved in (70 ml) of distilled water, and in order to obtain a homogeneous solution, it was placed in an ultrasonic device for (30 min).

Figure 1.
Pressure system is used in this work.

3.2 ZnS tablet preparation

Zinc sulfide (ZnS) powder was used after being pressed into tablets of a certain thickness, whereby (4 g) of the powder was weighed per tablet, then the powder was placed inside a stainless steel cylinder after cleaning it with ethanol, it was pressed using a compressive strength tester Type JB/T3818-1999), shown in **Figure 1**, with compressive strength (2 MPa).

4. Results and dissection

4.1 UV-visible absorption spectra measurements

The absorption spectrum of ultraviolet-visible (UV-Vi) rays was studied for graphene oxide, prepared by the modified Humer method and mentioned in paragraph (3). **Figure 2** shows the stages of formation of graphene oxide (GO). **Figure 2** shows

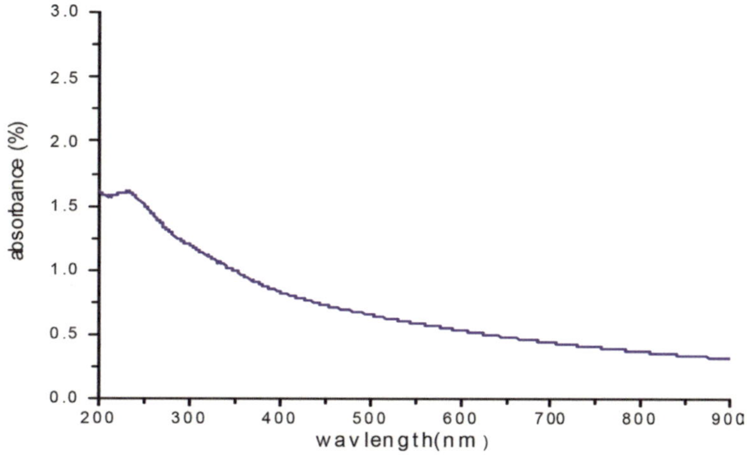

Figure 2.
Shows the UV-Vis absorption spectrum of graphene oxide.

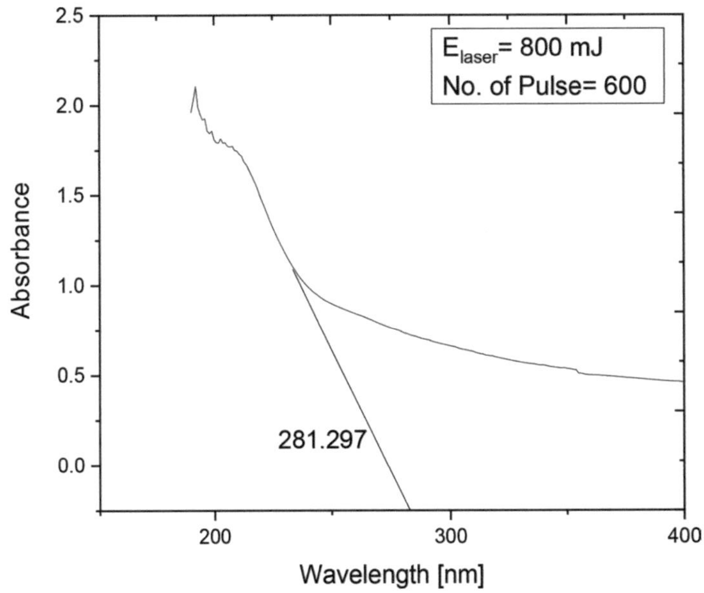

Figure 3.
Shows the Absorption Spectrum of sample (A).

the absorption spectrum of the aqueous solution of graphene oxide (GO), where it found an absorption peak due to GO at (230 nm) due to the electronic transition (π-π*), as a result of the presence of the (C-C) bond [3].

Also, the absorption spectra of ZnS/rGO prepared by the hydrothermal and pulsed laser ablation method were analyzed using a Nd-YAG laser (wavelength 1064 nm, frequency 6 Hz, and pulse width 10 ns) for two samples only, the first sample (A) was prepared By the method of pulsed laser ablation, with a capacity of (800 mJ) and the number of pulses (600 pulse), while the second sample (B) was prepared by the hydrothermal method. **Figures 3** and **4** show the absorption spectra of the prepared samples, and through the results for sample (A), the highest value of the adsorption of the mixed compound was determined at a peak of 281.497 nm, and this indicates the occurrence of transitions of carbon bonds (C-C). By studying on zinc sulfide, it has the highest absorption at 336 nm wavelength in most of the experiments, as well as for graphene oxide, it has the highest absorption at 230 nm [3].

In this work, it was found that the highest absorption of the mixture was at the wavelength of 281.497 nm, which indicates that there are spectral transitions that occurred between the π-π* transitions in the carbon (C-C) bond. The energy gap of the mixture was calculated using the method [4]. The Touc plot was about 4.942 eV, as shown in **Figure 5**.

Where the energy gap in most of the studies was 3.8 eV, as well as the wide band-gap 3.5 eV (3.8 eV), as well as zinc sulfide nanoparticles (ZnS-NPs) have recently received heightened thinking regarding their use in many applications [5].

The results for sample (A) indicate a shift blue spectral shift and this is consistent with the theoretical study indicating an increase in the energy gap of the synthesized ZnS nanoparticles as well as a slight amplification in the nanoparticles [6]. As for the results of the sample (B), the highest value of the adsorption of the compound was determined at a peak of 518.023 nm, and this indicates that transitions of carbon

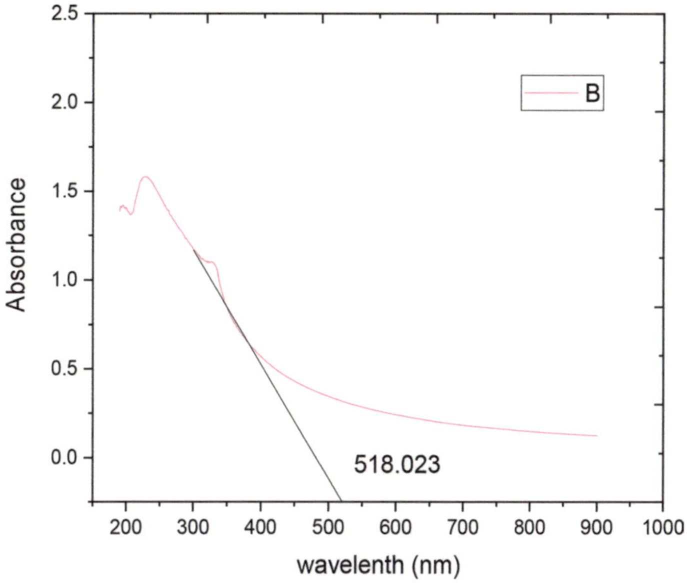

Figure 4.
Shows the Absorption Spectrum of sample (B).

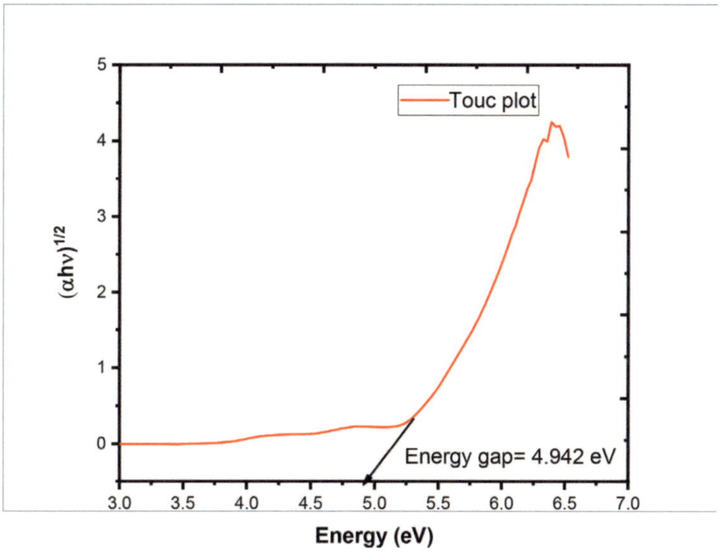

Figure 5.
Shows the energy gap of the sample (A).

(C-C) bonds also occurred. In this study, it was discovered that the mixture's peak absorption occurred at wavelength (281.297) nm, indicating that there were spectral transitions that took place in between transitions (-* in the C-C bond). In addition, the mixture's energy gap was determined using the Touc plot method and was approximately (2.983 eV), as shown in **Figure 6**. It is evident from the sample (B) results that there is a red shift, which results in a reduction in the energy gap of ZnS material.

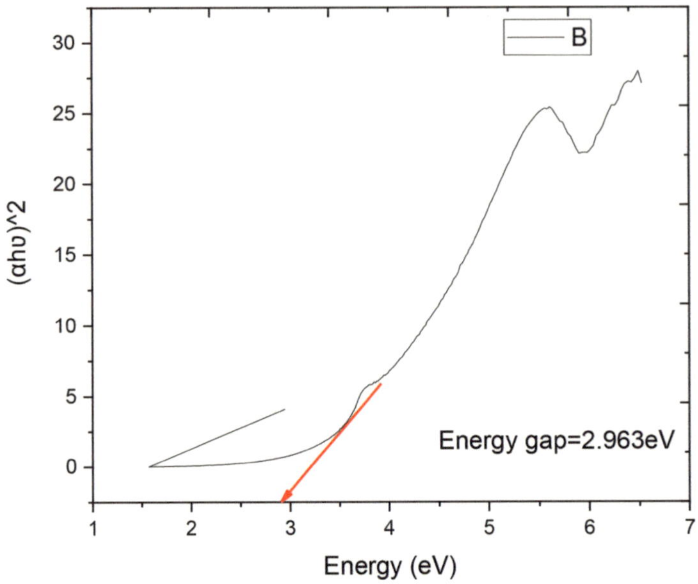

Figure 6.
Shows the energy gap of sample (B).

4.2 Results of X-ray diffraction (XRD) measurements

4.2.1 Results of X-Ray diffraction measurements

Using the X-ray diffraction spectrum for the purpose of studying the crystal structure of the samples prepared by the physical and chemical method. **Figure 7** shows the X-ray diffraction spectrum of the sample (A), where the (XRD) spectrum showed

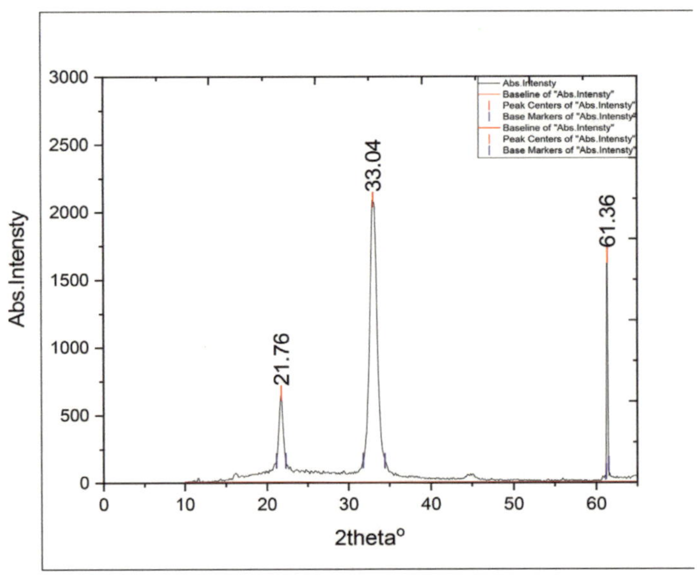

Figure 7.
Shows the (XRD) spectrum of the ZnS/rGo Nano composite prepared by the physical method.

a crystalline band at 2θ = 21.76°, 33.04°, 61.36°, The grain size was calculated using Scherer's equation after converting the angles from degrees to radians in addition to the values of the vertices.

The crystal size was calculated, and the results showed that the lowest value of the particle size was at the peak of 33.04°, and the diameter was about 8.359 nm, and this occurs due to the phenomenon of quantum confinement. Also, the average crystalline size of the particles was calculated and was about 25,233 nm, as shown in **Table 1**.

The X-ray diffraction spectrum of the sample (B), where the (XRD) spectrum showed a crystalline band at 2θ=(21.76, 33.04, 61.36) and the grain size was calculated using Scherrer equation after converting angles from degree scale to Radians plus vertices values.

Calculations of the crystal's size revealed that the lowest particle size value—33.04—was found at the peak, and the diameter—approximately 8.3596 (nm), which is caused by the phenomenon of quantum confinement. Additionally, the average particle size was determined to be around (25.23373) nm.

4.3 Results of Fourier Transform Infrared (FTIR) measurements

The FTIR spectrum was studied to determine the functional groups present in each of the two samples prepared by physical and chemical methods, They often interact with water layers and create a collection of chemical interactions. The FTIR spectra of ZnS/rGO for sample (A) is depicted in **Figure 8**. Experiments revealed that the peaks seen between 1636.01 and 3329.17 cm^{-1} are caused by the bending vibrations of adsorbed water molecules and the stretching vibrations of straight hydroxyl groups. Alkoxy and epoxy groups can also be found in the combination. The planar C=C band and the aromatic group vibrations in rGO [7] are what cause the peak to appear at 1636.01 cm^{-1}. However, they were marginally displaced to the lower frequency for ZnS-rGO at the mixed Nano composite [8].

Figure 9 shows the FTIR spectrum of ZnS/rGO for sample (B), through experimentation, it was shown that the peaks seen between 1636.12 and 3280.13 cm^{-1} are caused, respectively, by the stretching vibrations of the hydroxyl groups and the bending vibrations of the adsorbed water molecules. Alkoxy and epoxy groups are also present in the combination in the sample. The planar (C=C) band and the vibrations of the aromatic groups in rGO are responsible for the resultant peak at 1636.12 cm^{-1}. Notably, with the exception of being slightly displaced to the lower frequency, all of the aforementioned peaks were at ZnS-rGO (mixed Nano composite). This is explained by the chemicals' interaction with rGO and ZnS. In order to explain the absorptions at 559.81 and 408.54 cm^{-1}, the stretching of the vibrations of the ZnS bonds.

No.	Intensity	2 theta°	FWHM	Peak center	Crystallite size (D) nm	D Average (nm)
1	446.58	22.3	0.66757	21.76	12.11814	25.23373
2	2396.34	34.42	0.99123	33.04	8.359659	
3	292.11	61.48	0.16727	61.36	55.2234	

Table 1.
Shows the data of the XRD spectra of a compound for the sample (A).

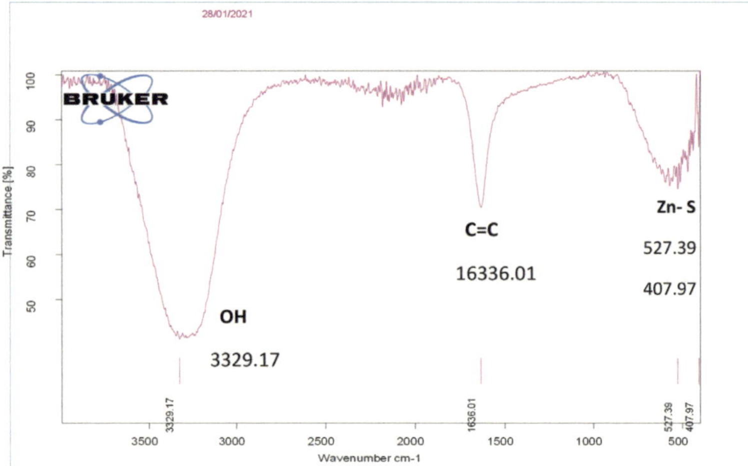

Figure 8.
Shows the FTIR spectrum of the sample (A).

4.4 Results of SEM and EDX scanning electron microscopy measurements SEM and EDXS analyses

Using a SEM device to identify the morphological characteristics of the prepared ZnS/rGO Nano composite samples, and by studying part of one of the images of the sample (A) at 200 nm in **Figure 9**, the results showed the formation of nanoparticles with different dimensions and measurements of their diameters, which were about 54 nm, in addition to the presence of particles with diameters Less, and this formation is due to the occurrence of quantum confinement phenomenon. Results of SEM measurements for samples (A,B) with distribution function of particles is given by **Figures 10–14**.

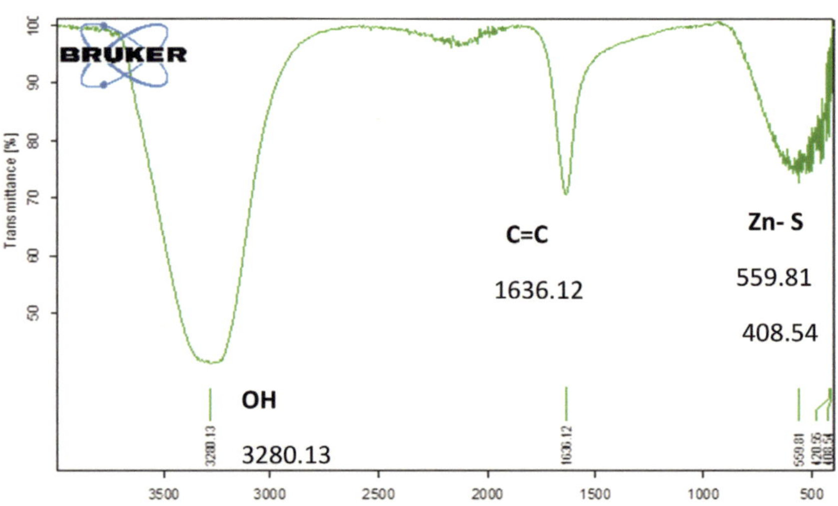

Figure 9.
Shows the FTIR spectrum of the sample (B).

Figure 10.
Shows the SEM images of the sample (A).

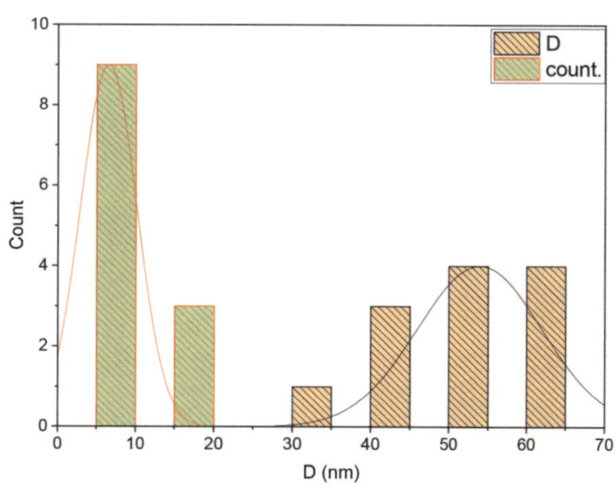

Figure 11.
Distribution function of particles of SEM in sample (A).

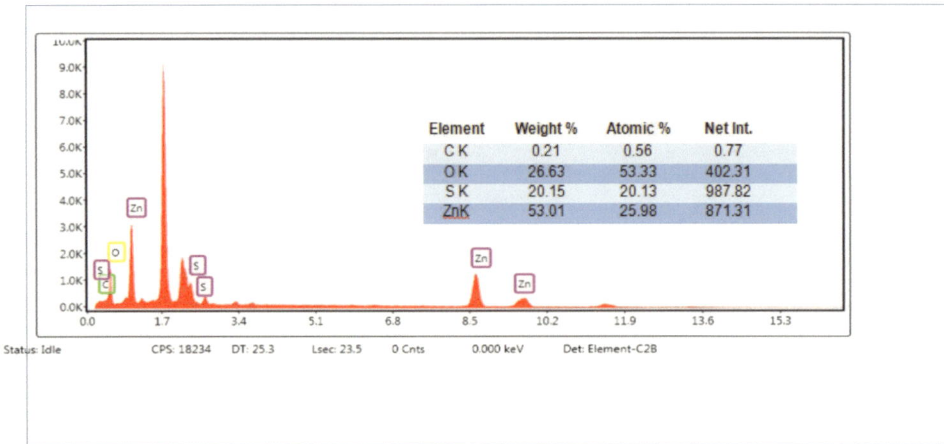

Figure 12.
EDAX for the ZnS/rGo complex with the element ratios of the sample (A).

Figure 13.
Shows the SEM images of the sample (B).

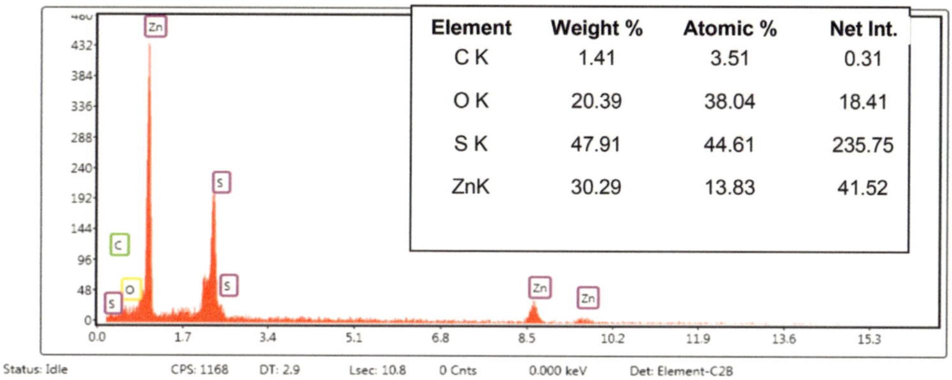

Figure 14.
Shows the distribution function for a portion of the SEM graph of the sample (B).

5. Application of removing dye from water using ZnS/rGO nano composite

The Nano composite ZnS/rGO prepared by the two methods was applied to remove the methylene blue dye (C16 H18 N3SCl. 3 H_2O) which is used in leather and textile coloring (4 mg) of the dye was dissolved in (100 ml) of distilled water and then (2 ml) was added From ZnS/rGO of sample (A) to (2 ml) of dye, then examining the samples by UV-visible spectroscopy. **Figure 15** shows the result of adding ZnS/rGO to sample (A) to the dye and its effect on removal. We note that the peaks for each (C16 H18 N3SCl. 3 H_2O + ZnS/rGO) from (5 min), (10 min) and (15 min) are almost similar and have lower adsorption intensity compared to ZnS/rGO. This indicates that The prepared ZnS/rGO Nano composite is capable of absorbing dyes and

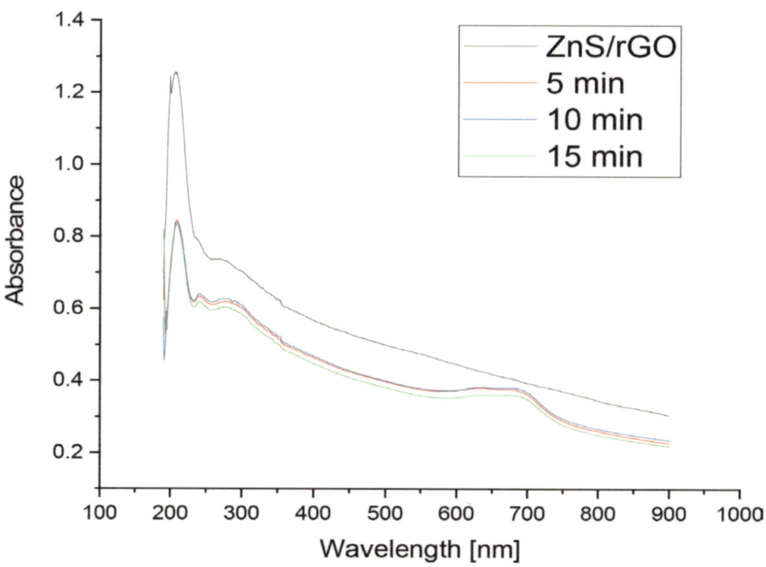

Figure 15.
Shows a UV-Vis spectrum showing the effect of the ZnS/rGO nano composite of the sample (A) on the removal of dye.

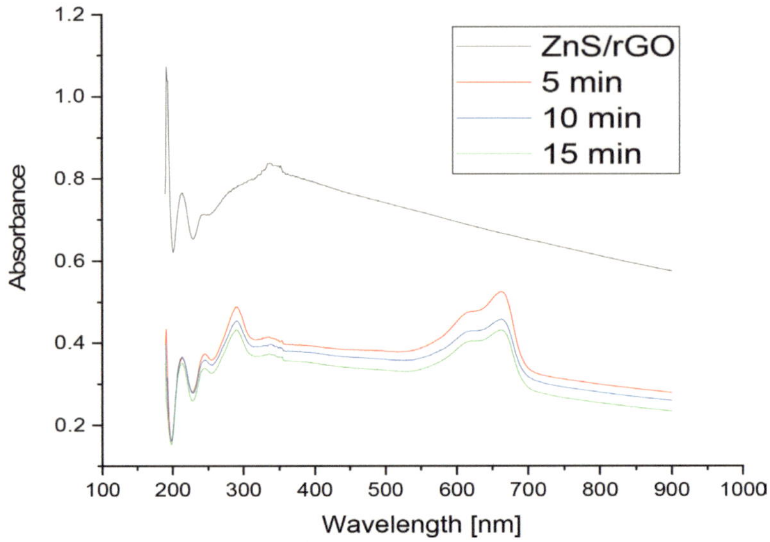

Figure 16.
Shows a UV-Vis spectrum showing the effect of the ZnS/rGO Nano composite of the sample (B) on the removal of dye.

removing pollutants from the water. Then (2ml) of ZnS/rGO was added to sample (B) to (2 ml) of the dye and then the samples were examined by UV-visible spectroscopy, **Figures 15** and **16** shows the result of adding ZnS/rGO to sample (B) to the dye and its effect on removal the peaks for each ((C16 H18 N3SCl. 3 H_2O + ZnS/rGO from (5 min) and (10 min) and (15 min) be satisfied It is almost opaque and has a lower adsorption intensity compared to ZnS/rGO. This indicates that the prepared ZnS/rGO Nano composite is capable of absorbing dyes and removing pollutants from water.

6. Conclusion

By studying pulsed laser ablation and comparing it with one of the chemical methods, it was found that it is the most suitable method for finding nanoparticles of small sizes due to the occurrence of the phenomenon of quantum confinement by the induction plasma. However, the density of nanoparticles is in a small proportion compared to chemical methods. Laser deamination is one of the important techniques in obtaining small nanoparticles with low densities, and this process is important in the manufacture of thin films as well as in the grafting process of some semiconductors used in electronic applications and detectors. The difference in the energy gap leads to the emergence of new physical properties.

Author details

Mohammed J. Jader* and Wasan H. Jawad
Department of Laser Physics, College of Science for Women, University of Babylon, Iraq

*Address all correspondence to: wsci.mohammed.jawad@uobabylon.edu.iq

References

[1] Buzea C, Pacheco II, Robbie K. Nanomaterials and nanoparticles: Sources and toxicity. Biointerphases. 2007;**2**:MR17-MR71

[2] Kim BH, Hackett MJ, Park J, Hyeon T. Synthesis, characterization, and application of ultrasmall nanoparticles. Chemistry of Materials. 2014;**26**:59-71

[3] Yang Y, Liu T. Fabrication and characterization of graphene oxide/zinc oxide nanorods hybrid. Applied Surface Science. 2011;**257**:8950-8954

[4] Sud A, Sharma RK. Study of optical properties of ZnS thin film using simulation software. International Journal of Applied and Natural Sciences (IJANS). 2016;**5**(5):13-18

[5] Tauc J. Amorphous and Liquid Semiconductors. New York: Plenum Press; 1974

[6] Fang X, Zhai T, Gautam UK, et al. ZnS nanostructures: from synthesis to applications. Progress in Materials Science. 2011;**56**(2):175-287

[7] Wang Z, Huang B, Dai Y, Zhang X, Qin X, Wang J, et al. Crystal Engineering Communication. 2012;**14**:1687

[8] Selorm E, Messai A, Mamo A. Cobalt-doped ZnS-reduced graphene oxide nanocomposite as an advanced photocatalytic material. Porous Materials. 2015;**22**:47-56

Section 2

Modeling

Chapter 4

Bases of Mathematical Simulation of Modes Obtaining Ultrafine Powder by Laser Ablation of Metal Particles Vertically Falling under the Action of Gravity in an Inert Gas Medium

A.Ph. Ilyushchanka

Abstract

The fundamentals of mathematical modeling of the modes of obtaining an ultrafine powder by metal particles laser ablation of vertically falling in an inert gas medium under the action of gravity, provided that these particles flow around the laminar flows of this gas and are outlined. Using the results of the research as an example, the practical possibility of using these foundations of mathematical modeling is shown when obtaining ultrafine powders with particles having a shape close to spherical by the proposed method of laser ablation from spherical titanium particles of a fraction of 45–63 μm falling vertically in an inert gas medium. It is recommended to use the developed foundations of mathematical modeling in the design and development of laser ablation-based technologies for obtaining ultrafine and nanosized powders, including those in demand in the additive manufacturing of powders.

Keywords: laser ablation, modeling and simulation, metal powder, spherical particle, laser conical beam, laser flux

1. Introduction

It is known [1] that due to the high chemical activity and lower sintering temperatures, ultradispersed (less than 1 μm) powders, which may include nanosized (less than 100 nm), as a feedstock are increasingly used in existing and newly developed emerging technologies for obtaining promising materials with unique and enhanced physical and mechanical properties, which are in demand not only in mechanical engineering and medicine, but also in the chemical industry and energy. One of the most frequently used and intensively developing methods for obtaining ultrafine powders in powder metallurgy is laser ablation of solids. It differs from other known methods in its relative ease of implementation and the absence of accompanying impurities, which inevitably appear when similar powders are obtained by chemical methods or high-energy grinding. Currently, laser ablation is understood not just as the process of a possible transition of a substance from a solid state to a gaseous state under the action of monochromatic coherent electromagnetic radiation in the wavelength range from near infrared to red and green. At the same time, the process of transition of a substance in a vapor–gas state to its solid-state state is also understood. In addition to the vacuum induction gas atomization method, the laser ablation method makes it possible to obtain ultrafine powders with particles that have a shape close to spherical. Ultrafine metal powders with such particles are a promising feedstock for developing additive manufacturing [1–3]. As a rule, in order to obtain ultrafine powders, one tends to use laser ablation on solid, immobile, evaporating targets. Another similar method [4] based on metal particles falling vertically in an inert gas medium under the action of its own gravity may also be promising and relatively easy to implement for obtaining such powders. The development of the theoretical foundations of mathematical necessary modes for obtaining ultrafine powders modeling under such conditions can be facilitated by the formulation of the corresponding problem with an indication of the way to solve it. The formulation of such a problem, indicating the way to solve it using an illustrative example, is the main goal of this work. At the same time, the purpose of the work is also laboratory research, on the example of which it is possible to show the possibility, on the one hand, of performing mathematical modeling of laser ablation modes and obtaining ultrafine powder from spherical particles of titanium with a fraction of 45–63 μm, and on the other hand, reflects the pattern the course of this process. It should be noted that the development of the proposed method of laser ablation with metal particles sheerly falling in the field of gravity can contribute to the improvement of technologies for obtaining, using laser ablation, not only ultrafine and nanosized powder particles that are promising for additive production, but also purification from surface impurities and spheroidization, as themselves, particles with a fraction of less than 100 μm, and their laser selective layer-by-layer surfacing used in 3D printers, which are structurally similar to DMT (Direct Metal Tooling) 3D printers of the South Korean international company InssTex MX.

2. Problem statement, determination of the movement nature, and modes of particles of metal powder supply to the laser ablation zone

Let metal powder particles have a monospherical shape and a smooth surface, and a metered feed device provides them with a consistent exit and a vertical fall without collisions in an inert gas environment. In this case, they fall inside a hollow optically transparent tube (**Figure 1**) with smooth inner surfaces. The inside tube diameter (D_T) is larger than the particle diameter (D_0). Then, the falling of particles under these conditions will be prevented by the emerging forces of buoyancy (F_A) and gas-dynamic resistance (F_r) [5, 6]:

$$m_{pp}a = \gamma_{pp}\frac{\pi \cdot D_0^3}{6}\frac{dv}{dt} = m_{pp}g - F_A - F_r = \left(\gamma_{pp} - \gamma_g\right) \cdot g \cdot \frac{\pi \cdot D_0^3}{6} - F_r, \qquad (1)$$

where m_{pp}, γ_{pp}—mass and density of material (metal) particles; γ_g—gas density; v, t, a, g—speed, time, acceleration, and gravitational acceleration.

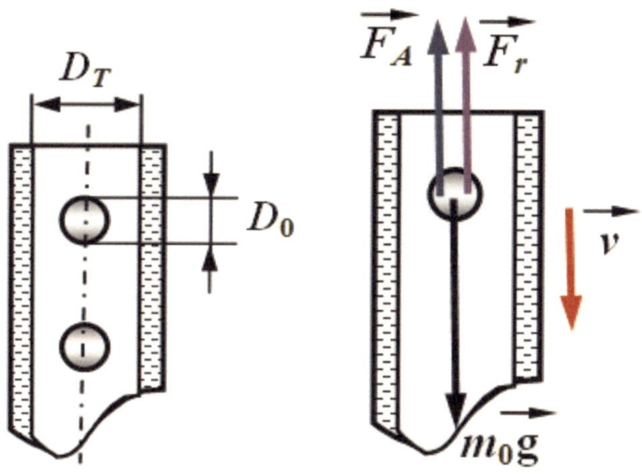

Figure 1.
Forces acting on particles when falling in an inert gas.

The drag force F_r in expression (1) is influenced not only by the shape, size, and speed acquired by the particles, but also by the density and viscosity (η) of the inert gas, including the drag coefficient $C_x = C_x(\text{Re})$, which depends on the Reynolds number [5, 6]:

$$\text{Re} \leq 0,1 : F_r = 3 \cdot \pi \cdot \eta \cdot D_0 \cdot v, \tag{2}$$

$$0,1 < \text{Re} \ : F_r = \frac{C_x \cdot \pi \cdot D_0^2 \cdot \gamma_g \cdot v^2}{8}, \tag{3}$$

$$0,1 < \text{Re} \leq 2 : C_x = \frac{24}{\text{Re}}; \ 2 < \text{Re} \leq 700 : C_x = \frac{24}{\text{Re}} + \frac{4}{\sqrt[3]{\text{Re}}}; \ 700 < \text{Re} : \ C_x = 0,4\text{–}0,5. \tag{4}$$

and besides:

$$\text{Re} = \frac{v \cdot \gamma_g \cdot D_0}{\eta}. \tag{5}$$

Depending on the Reynolds number, most often [5, 6] the following types of motion of gas flows flowing around the surfaces of moving bodies are distinguished: laminar (Re < 0.1), close to it (0.1 < Re < 2), and transitional to turbulent (2 < Re < 700) and turbulent motion (700 < Re). Of these types, for free-falling particles of metal powders with fractions D_0 = 40–150 μm and D_0 = 10–40 μm, the most characteristic is laminar or close to it motion, in which Re < 2. Therefore, the expression of resistance forces (3) is reduced to its particular representation (2), and the limiting values for particle velocities, at which their motion can still be considered close to laminar, are determined by the expression:

$$v_{\max} = \frac{2 \cdot \eta}{\gamma_g \cdot D_0}. \tag{6}$$

It is noted that in natural conditions inert gases behave as ideal gases. Therefore, their density and dynamic viscosity will depend both on the mass (m_{0g}), diameter (d_{0g}), effective cross section (σ), most probable velocity (v_{0g}) and mean free path (λ) of molecules between collisions, and their volume concentration (n) [7]:

$$\eta \sim \frac{\lambda_{0g} \cdot v_{0g} \cdot \gamma_g}{3}; \lambda_{0g} = \frac{1}{n \cdot \sigma \cdot \sqrt{2}}; \sigma = \pi \cdot d_{0g}^2 \cdot \left(1 + \frac{T_S}{T}\right); v_{0g} = \sqrt{\frac{8 \cdot k_B \cdot T}{\pi \cdot m_{0g}}}; \quad (7)$$

where k_B—Boltzmann's constant; T_S—Sutherland constant.

In contrast to the density ($\gamma_g = n{\cdot}m_{0g}$) and pressure ($p = n{\cdot}k_B{\cdot}T$) of an inert gas, which directly depend on the volume concentration (n) of molecules, its dynamic viscosity, as followed from expressions (7), will depend only on temperature [7]:

$$\eta(T) = \eta_0(T_0) \cdot \frac{T_0 + T_S}{T + T_S} \cdot \left(\frac{T}{T_0}\right)^{\frac{3}{2}}. \quad (8)$$

If the free fall of particles in an inert gas is not limited by height, then their speed will increase along with the drag force. This, after a certain time (τ_{pp}), will lead to the appearance of a certain steady rate of fall (v_{pp}), after which they will acquire a suspended state. The transition to this state is determined by the values of τ_{pp} and v_{pp}, which are obtained based on the solution of expression (1):

$$v_{pp} = \frac{\left(\gamma_{pp} - \gamma_g\right) \cdot g \cdot D_0^2}{18 \cdot \eta \cdot \chi_D}; \tau_{pp} = \frac{\gamma_{pp} \cdot D_0^2}{18 \cdot \eta}, \quad (9)$$

where $\chi_D = 1 + \frac{2,4 \cdot D_0}{D_T}$.

Let us take the above as the condition of the problem, using the MathCAD-15 application package and known data [8, 9] for titanium, we will study the nature of the motion of vertically falling particles with a diameter D_0 = 50 μm of the powder of this metal in an argon medium with a constant pressure p = 100 KPa depending on from changes in its temperature within T = 0–500°C. Calculation studies have established that an increase in temperature leads to a nonlinear decrease in the Reynolds number, which can have values close to Re = 0.04 at T = 0°C (273 K) and Re = 0.006 at T = 500°C (773 K), and in the subrange T = 300–500°C (**Figure 2a**)—0.04 and 0.006, respectively.

The values of the steady velocity, as well as the Reynolds number, will decrease monotonically with increasing temperature. In particular, at T = 0°C, they can be close to v_{pp} = 0.2 m/s, and at T = 500°C–v_{pp} = 0.09 m/s, and in the subrange T = 300–500°C (**Figure 2b**)–v_{pp} = 0.11 m/s and v_{pp} = 0.09 m/s, respectively.

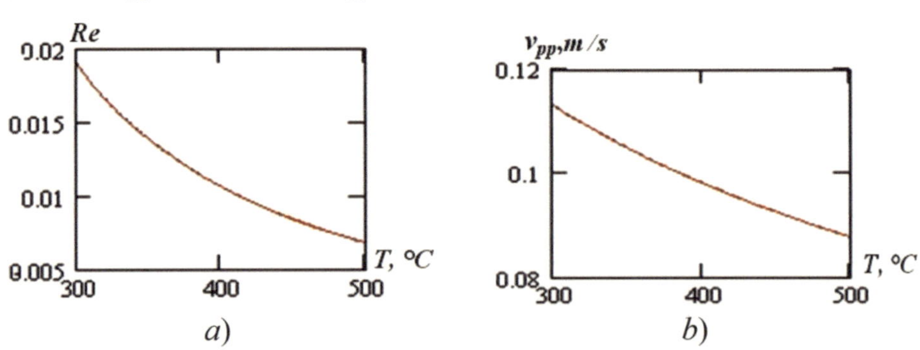

Figure 2.
The values of the Reynolds number (a) and the steady velocity of particles (b) depending on temperature.

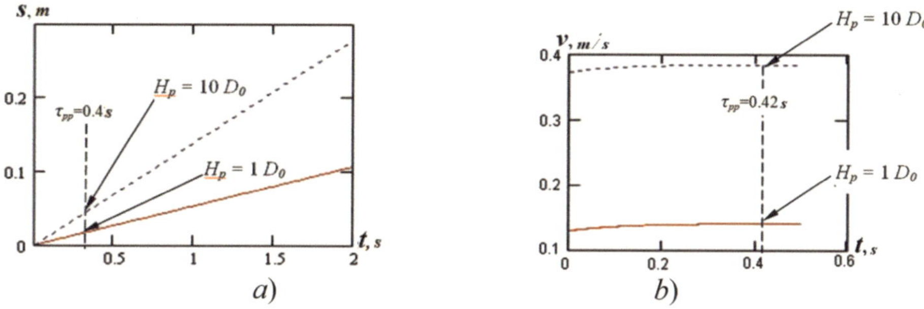

Figure 3.
The distance traveled (a) and the time of acquisition of a steady velocity (b) by particles.

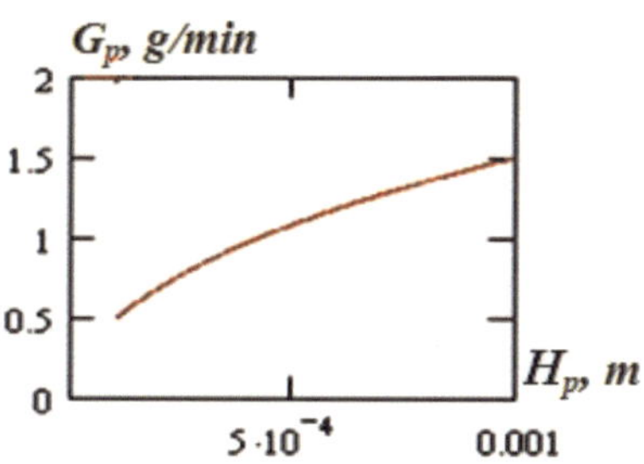

Figure 4.
Consumption of precipitated particles depending on the height of the titanium powder layer.

Figure 3 shows the results of modeling the path traveled and the time it took particles to acquire a steady velocity for the case when, during precipitation, they experienced pressure from the overlying layers of other particles, the height (H_p) of which was $H_p = D_0$ and $H_p = 10{\cdot}D_0$. The changes in the consumption of powder (G_p, g/min) of titanium and the amount of ejected particles associated with it are shown in **Figure 4**.

The nature analysis of the change in the steady velocity of powder particles (titanium) falling vertically in an inert gas (argon) with increasing temperature (**Figure 2**) decreases and in the studied temperature range can differ by at least an order of magnitude (v_{pp} = 0.18 m/s at T = 0°C and v_{pp} = 0.09 m/s at T = 500°C). The time taken for the particle velocity to reach stable values depends on the pressure and temperature of the gas, but does not depend on the pressure of the powder layers pushing these particles out (**Figure 3**). It also follows from computational studies that in the temperature range T = 0–500°C, the nature of motion corresponding to laminar or close to it, for all incident particles of 10 μm < D_0 < 100 μm of titanium powder in an inert gas, will be preserved. If we single out a subrange within T = 300–500°C from this temperature range, then for particles D_0 < 200 μm. In this temperature subrange, the nature of motion corresponding to close to laminar motion will be retained for particles with D_0 < 100 μm even if the pressure in argon increases to 300–500 kPa.

If (**Figure 5**) titanium particles falling in a hollow cylindrical tube 1 acquire a stable velocity into the zone of action of conical laser beams 2, which has a length of l_{cb} = 10 mm, then the number of particles (N_{pp}) squeezed out by a layer of powder with a height of $H_p = 10{\cdot}D_0$ and $Hp = 10{\cdot}D_0$ in the indicated zone will correspond to the data given in **Table 1** and in **Figure 6**.

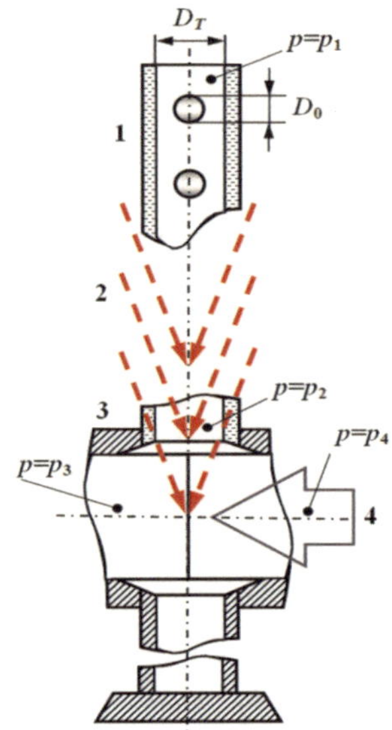

Figure 5.
Scheme of laser ablation of powder particles. 1: Hollow optically transparent tube; 2: Laser radiation conical beam; 3: Laser ablation chamber; 4: Inert gas flow.

G, g/min	N_{pp}, p.	v_{pp}, m/s	t, s
0,5	2003	0,13	0,2
1,5	2146	0,38	0,07

Table 1.
The quantity, speed and time spent by particles in the laser action zone.

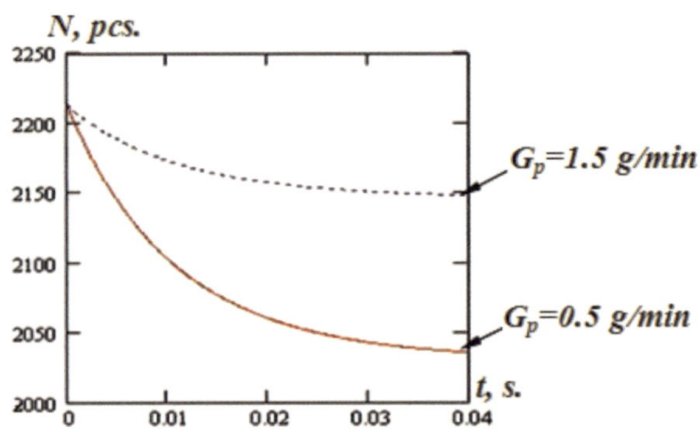

Figure 6.
The dependence of the particles number on the time of transition to a suspended state.

When the inert gas flow is set in the laser ablation chamber 3 (**Figure 5**), a pressure difference (Δp) will appear at the exit of the hollow optically transparent tube 2, which can create an additional force that contributes to the deceleration of the falling particles. Taking this into account and taking into account expression (1), it is possible to purposefully carry out additional adjustment of the steady velocity of these particles in the area of action of the conical laser beam:

$$v_{pp} = \frac{\left(\gamma_{pp} - \gamma_g\right) \cdot g \cdot D_0^2}{18 \cdot \eta \cdot \chi_D} - \frac{\Delta p \cdot D_0}{12 \cdot \eta}. \tag{10}$$

With a laminar flow of an inert gas through a laser ablation chamber 3 (**Figure 5**) with smooth walls, based on the laws of continuity and conservation of energy, such adjustment can be made using a pressure drop (Δp):

$$\Delta p = p_2 - p_1 = \frac{p_4 \cdot G_{g4} - p_3 \cdot G_{g3}}{G_{g2}} - p_1, \tag{11}$$

where p_1–p_4—pressure at the inlet of a hollow optically transparent tube, in the area of action of the conical laser beam, at the outlet and inlet of the laser ablation chamber; G_{g2}–G_{g4}—volume flow rate of gas at the exit of the zone of action of the conical laser beam, at the exit and inlet of the laser ablation chamber.

On the example of studying the nature of the motion of titanium powder particles in an inert gas, limited by the inner diameter of a hollow optically transparent tube, and the results obtained, the following conclusions can be drawn.

1. In a free vertical fall under the action of gravity, the particles of the medium and fine fractions of metal powders located in an inert gas experience deceleration due to the emerging buoyancy forces and gas-dynamic resistance, which affects the nature of their movement. A feature of this movement is that the gas flows around the particles are essentially laminar or close to laminar. The speed of the particles, which increases with the fall, after a certain period of time, reaches the stable values.

2. The path traveled, the time of transition to a constant value of the falling velocity of particles, and the value of this velocity itself are influenced not only by the size of the particles, but also by the density, pressure, and temperature of the inert gas in which they move. If argon is an inert gas, then in the selected range of its temperatures T = 0–500°C and pressures p = 100–500 kPa, a change in the first parameter (temperature) will, with a high degree of certainty, lead to an inversely proportional change in the value of the steady velocity, and a change the second parameter (pressure), on the contrary, to a linear change in this value.

3. The steady velocity of the particles is also affected by the pressure of their ejection from the metered feeder. If the particles move in a hollow optically transparent tube into the zone of action of the conical laser beam, and from it into the laser ablation chamber, then it is possible to additionally adjust the time of their transition to the equilibrium state and the value of the steady velocity by varying the pressure, temperature, and flow rate of the supplied inert gas. These parameters, including the granulometric composition of the supplied powder and its flow rate, as well as the pressure that pushes out the particles, all these taken together determine the modes of powder supply to the laser ablation zone.

3. Conical laser beams effect on vertically falling particles of metal powders, and determination of modes for obtaining finer fraction powders from them

Let us assume (**Figure 5**) that conical laser beams intersect along the axis of a hollow cylindrical tube on a segment of a certain length (l_{cb}) and the total laser radiation flux formed on this segment is uniform. Then, the density (q_{cb}) of the flow when falling vertically and successively falling particles of metal powders will be determined by the expression:

$$q_{cb} = \frac{P}{S} = \frac{P}{\pi \cdot D_0 \cdot l_{cb}} = \frac{1}{S} \cdot \frac{dW}{dt} = \frac{W_{puls}}{S \cdot \tau_{puls}}; q_{\Phi} = q_{cb} \cdot \frac{l_{cb}}{D_0}, \quad (12)$$

where P, W, W_{puls}, τ_{puls}—power, energy, pulsed energy, and pulse duration of laser radiation; S—the area of laser action on particles formed by conical beams; q_{Φ}—flux density of laser radiation projected onto the particle surface.

In this case, the density (q_0) of the laser radiation flux absorbed by the surface layer of particles and the light pressure exerted by it (p_{red}) will be determined based on the Bouguer (Beer–Lambert) law [4, 7]:

$$q_0(\xi) = q_{\Phi} \cdot (1 - R_{ref}) \cdot \exp\left(-\int_0^{\xi} \chi(\xi) d\xi\right); p_{red} = \frac{2 \cdot R_{ref} \cdot q_{\Phi}}{c} \cdot \exp\left(-\int_0^{\xi} \chi(\xi) d\xi\right) \quad (13)$$

where ξ—depth of penetration of laser radiation into the metal surface; R_{ref}—reflection coefficient; χ—linear absorption coefficient; c—light speed.

If the particles surfaces absorbed by the laser beam have a juvenile metal surface, then χ = Const, then expressions (13) will take a simpler form:

$$q_0(\xi) = q_{\Phi} \cdot (1 - R_{ref}) \cdot \exp(-\chi \cdot \xi); p_{red} = \frac{2 \cdot R_{ref} \cdot q_{\Phi}}{c} \cdot \exp\left(-\int_0^{\xi} \chi \cdot d\xi\right). \quad (14)$$

Let the laser radiation flux be uniform in the zone of action of the conical beam. Then, in accordance with expressions (13) and (14), in the surface layer of particles to a depth ($\xi \sim 10^{-8}$–10^{-7} m) of penetration of laser radiation into the metal,a heat source arises that generates a heat flux having a uniform density (q_0). This heat flux during the action of the laser radiation pulse can lead to the detachment (ablation) of some atoms from the particle surface by the evaporation mechanism and heat the material of the particle itself. The relationship between the heat flux density generated by a surface heat source and the heat flux densities causing ablation (q_{abl}) and heating (q_{λ}) is established by the heat balance equation: $q_0 = q_{abl} + q_{\lambda}$. Then, according to Refs. [4, 7], the thermal flows causing ablation (q_{abl}) and heating (q_{λ}) are determined by known transcendental equations:

$$q_{abl} = \frac{(H_{м} + 2,5 \cdot R_g \cdot T_{\chi}) \cdot \sqrt{\gamma_{pp} \cdot E_M}}{\mu \cdot \exp\left(\frac{H_{м}}{R_g \cdot T_{\chi}}\right)}, \quad (15)$$

$$q_{\lambda} = \pi \cdot (T_{\chi} - T_0) \cdot \gamma_{pp} \cdot c_{pp} \cdot D_0 \cdot f_{\Phi} \cdot \frac{B_1(Pe)}{B_0(Pe)}, \quad (16)$$

where $H_м$—molar enthalpy of phase transition of a metal into a gaseous state; R_g—universal gas constant; T_0, T_χ,—initial temperature and ablation temperature; E_M—elastic modulus; μ—molar mass; f_Φ—laser pulse frequency; B_0, B_1—zero and first-order Bessel functions; Pe—Peclet number.

The temperature distribution inside the particle material when heated by a heat flux formed by laser action is determined by a well-known equation, which in differential form has the following form:

$$\frac{\partial T}{\partial t} - a_{pp} \cdot \Delta_L(T) = \frac{q_{\lambda V}}{\gamma_{pp} \cdot c_{pp}} \approx \frac{q_\lambda}{\gamma_{pp} \cdot c_{pp} \cdot \Delta h}, \tag{17}$$

where a_{pp}—coefficient of thermal diffusivity of the particle material; $q_{\lambda V}$—specific thermal power; Δ_L—Laplace operator; c_{pp}—heat capacity of the particle material; Δh—heating thickness of the surface layer during the time (τ_{puls}) of the action of the laser pulse, and:

$$a_{pp} = \frac{\lambda_{pp}}{c_{pp} \cdot \gamma_{pp}}, \Delta h = \sqrt{a_{pp} \cdot \tau_{puls}},$$

λ_{pp}—thermal conductivity coefficient of the particle material.

If we assume that laser ablation is associated with the instantaneous evaporation of material from the particle surface, then the ablation pressure (p_{abl}) will counteract the surface tension pressure (p_L). These pressures are at least an order of magnitude higher than the static and dynamic pressures of the surrounding gaseous medium [7]. Therefore, we will assume that the resulting pressure drop $\Delta p = p_L – p_{abl}$ and the volume of material removed by ablation (ΔV_{pp}) of particles are interconnected according to the Claperon-Clausius law [7], which in differential form has the form:

$$\frac{\partial p}{\partial T} = \frac{H_м}{T \cdot \Delta V_{pp}} \approx \frac{\Delta p}{T_\chi - T_0} = \frac{H_м}{T_\chi \cdot \Delta V_{pp}}. \tag{18}$$

If we assume that $\Delta p = p_0$, and the molar enthalpy of the phase transition of the metal to the gaseous state is a constant value, then expressions (19) are transformed into the known transcendental equation:

$$\ln\left(\frac{p_0}{p}\right) = \frac{H_м}{R_g} \cdot \left(\frac{1}{T_\chi} - \frac{1}{T_0}\right), \text{or} p(T) = p_0 \cdot \exp\left(-\frac{H_м}{R_g} \cdot \left(\frac{1}{T_\chi} - \frac{1}{T_0}\right)\right) \tag{19}$$

Eq. (19) reflects the change in the vapor pressure formed by the evaporated metal, and the velocity (v_{mg}) of movement of metal atoms forming vapors is the Hertz-Knudsen equation:

$$v_{mg}(T) = \frac{0,82 \cdot p_0}{\gamma_{pp}} \cdot \left(\frac{\mathcal{M}_{pp}}{R_g \cdot T_\chi}\right)^{\frac{1}{2}} \cdot \exp\left(-\frac{H_м}{R_g} \cdot \left(\frac{1}{T_\chi} - \frac{1}{T_0}\right)\right) \tag{20}$$

Visually, laser ablation is accompanied by a rapidly expanding torch filled with vapor-forming metal atoms. If its expansion is taken as free self-similar, characteristic of a polytropic ideal gas [7] with a known polytropic index (ζ_p), then, as follows from [4, 7], the maximum velocity (v_F) of the expanding flame front will be determined by the expression:

$$v_F(\max) = \frac{2}{\zeta_p - 1} \cdot \sqrt{\frac{\zeta_p \cdot R_g \cdot T_\chi}{\pi \cdot \gamma_{pp} \cdot D_0^2 \cdot \Delta h}}. \quad (21)$$

The flame removed from the surface of the particle, due to the high speed of the front and rapid expansion, will not fully carry out heat exchange with the environment, which is unlimited in relation to it, practically immobile with pressure and temperature that are not significantly different from normal ones. This allows laser ablation and the rapidly expanding torch accompanying it to be considered as interconnected isolated thermodynamic systems with the polytropic index $\zeta_p = 5/3$, to which the superposition principle is applicable. Then, during laser ablation of particles and flame expansion, evaporation and condensation of metal droplets occur, which is determined by expressions (18) and (19). In this case, droplet condensation occurs due to the removal of a heat flux of uniform density ($q_{v\text{-}l}$) at the "saturated vapor-droplet surface" interface in a known way:

$$q_v = \alpha_v \cdot (T_v - T_s), \alpha_v = \frac{2 \cdot B_l \cdot c_d}{D_d} \cdot \left(\frac{\gamma_d^4}{\mu}\right)^{\frac{1}{3}} \quad (22)$$

where α_v—heat transfer coefficient; T_v, T_s—vapor and drop surface temperatures; B_l—proportionality coefficient to the velocity of propagation of elastic waves in liquid metal; D_d—drop diameter; γ_d—density of liquid metal in a drop.

Taking into account the results of metal sputtering presented in [1, 10, 11], as well as the above assumptions and dependences (19) and (20), the dimensions (D_p) of new powder particles obtained from condensed metal drops will largely correspond to the normal logarithmic distribution $F(D_p)$:

$$F(D_p) = \frac{1}{D_p \cdot \sigma_p \cdot \sqrt{2\pi}} \cdot \exp\left(-\frac{\left[\ln(D_p) - \ln(D_{p0})\right]^2}{2 \cdot \sigma_p{}^2}\right), \quad (23)$$

$$\sigma_p = \sqrt{\frac{1}{n-1} \cdot \sum_{i=1}^{n} \left[\ln(D_{pi}) - \ln(D_{pp})\right]^2}, \; D_{p0} = \frac{1}{n-1} \cdot \sum_{i=1}^{\infty} (D_{pi}), \quad (24)$$

where σ_p—dispersion; D_{p0}—average or expected particle diameter.

If necessary, D_{p0} can also be determined taking into account expression (22):

$$D_{p0} = \frac{2 \cdot B_l \cdot c_d}{r_v} \cdot \left(\frac{\gamma_d}{\mu}\right)^{\frac{1}{3}} \cdot \left(T_\chi - T_l\right) \cdot \sqrt{\frac{\pi \cdot \mu_{pp}}{8 \cdot R_g \cdot T_l}}, \quad (25)$$

where r_v—vaporization specific heat.

Taking the above as the next condition of the problem and using the application package MathCAD-15, as well as the known data [8, 9] for titanium, we study the course of laser ablation under the action of conical beams (**Figure 5**) on powder particles, taking into account the data of **Tables 1** and **2**. If, in this case, we use not

Pulse duration, ms	Pulse repetition rate, Hz	Pulse energy, J	Radiation wavelength, nm
3	50	up to 3	1060

Table 2.
Characteristics of the pulsed laser radiation source.

DOI: http://dx.doi.org/10.5772/intechopen.111672

only dependences (14)–(17), but also take into account the method of heat sources, as well as the boundary conditions of the first and third kind [7, 12], then the temperature distribution inside the particle at the initial moment of exposure to the laser beam radiation will have the form shown in **Figure 7**.

When using the temperature distribution inside the particle obtained by the method of heat sources as a calibration function, then the temperature spread obtained taking into account the boundary conditions of the first and second kind will look as shown in **Figure 8**.

When conducting virtual studies, it was taken into account that ablation from the surface of particles occurs at temperatures significantly higher than the critical Debye temperature; therefore, the electronic and atomic thermal conductivities (**Figure 9a**) were reduced to an average value that determines the thermal diffusivity as an inseparable concept (**Figure 9b**).

At the same time, the effect of the thickness of powder pressure layer on the surface temperature of vertical falling particles and its distribution in the surface layer at the depth of r = 5 μm after the effect of a single pulse of laser radiation was taken

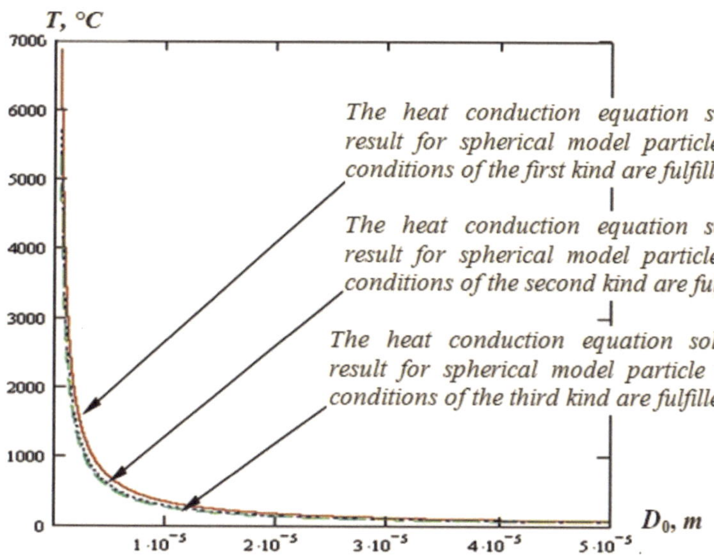

Figure 7.
The temperature inside the particle at the initial moment of exposure to the laser radiation flux.

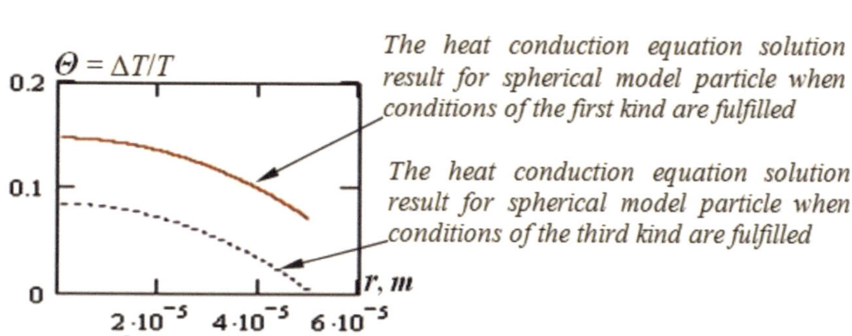

Figure 8.
Possible spread of temperature inside the particle at the initial moment of exposure to the laser radiation flux.

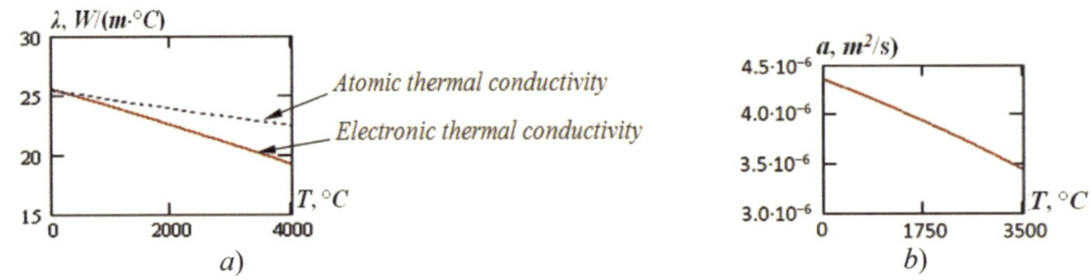

Figure 9.
Temperature effect on the thermal conductivity (a) and thermal diffusivity (b) of titanium.

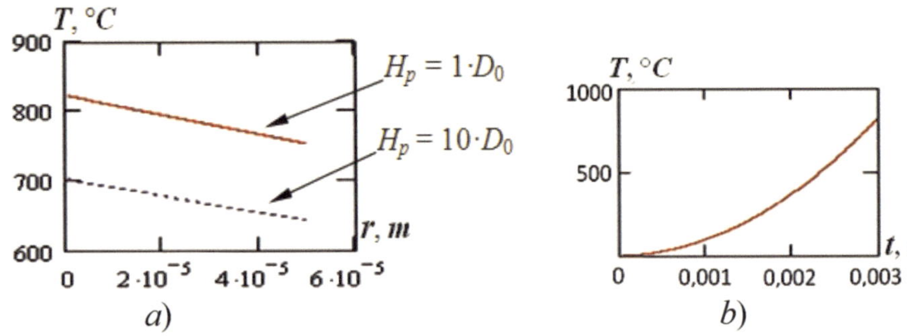

Figure 10.
Temperature change inside (a) and at the boundary of the near-surface layer (b) of the particle under the influence of a single pulse of laser radiation.

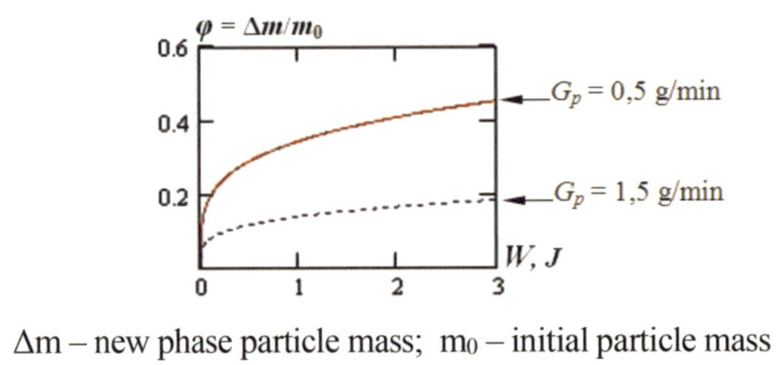

Figure 11.
Influence of the energy of a single pulse of laser radiation on the relative ablation.

into account (**Figure 10a**). The character of variation of particle surface heating temperature from this pulse is shown in **Figure 10b**.

Based on **Figures 6–10** and **Table 1**, it follows that the value of the relative ablation of the material from the particle surface (**Figure 11**) will depend on the amount of powder supplied to the laser impact zone (on its consumption). If the energy of the laser radiation pulse is $W = 3$ J, then at a flow rate $G = 0.5$ g/min, it will be 45–47% of the entire mass of the particle, and at a flow rate of the studied powder $G = 2$ g/min it will be only 16–18%.

Based on those shown in **Figures 7–11** results, it is assumed that at normal ($p = 100$ kPa) pressure of an inert gas (argon), a complete transition to the vapor–gas phase of titanium particles with a diameter of $D_x = 50$ μm that fell into the laser ablation zone (**Figure 5**) can occur at a speed powder supply of G = 0.2 g/min., 50%—at G = 0.5 g/min., and 20%—at G = 1.8 g/min. At this flow rate, the granulometric distribution $F(D_x)$ of the condensed particles of the new phase will be as shown in **Figure 12**. If at G = 0.2 g/min. Stepwise increases the pressure of the inert gas, then the granulometric distribution will have the form shown in **Figure 13**, and the relative mass distribution (M) will be as shown in **Figure 14**.

Of those presented in **Figures 12** and **13** results of the normal-logarithmic granulometric distribution, it follows that in the case of laser ablation under consideration, the condensed particles of the new phase form a powder, the fractional composition of which will be fifty times less (~ 1 μm) than the original one (~D_0 = 50 μm). If an ideal gas filter from dust with particles smaller than 250 nm is placed at the exit of the laser ablation chamber (**Figure 5**), then after it the granulometric distribution of condensed particles of the new phase will be as shown in **Figures 15** and **16**.

From the presented **Figures 14–16** of the data it follows that within an hour at G = 1.8 g/min, nine times more particles of the initial titanium powder will pass through the laser ablation zone than at a flow rate of 0.2 g/min. In this case (**Figure 17**), the number of particles of a new phase formed after laser ablation with a size of less than 250 nm will also be nine times greater.

Analysis of the model studies results (**Figures 12–17**) allows us to draw the following conclusions.

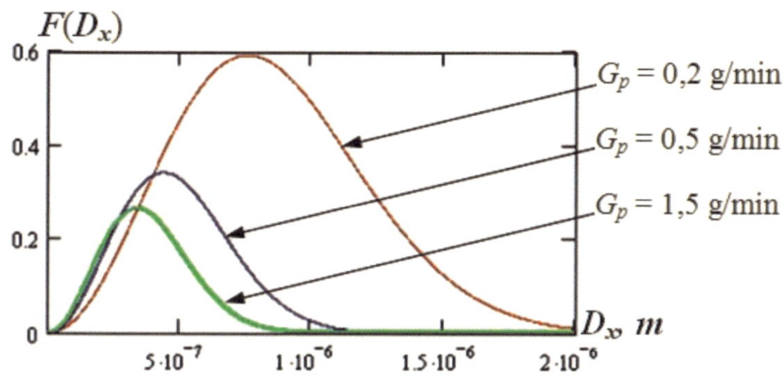

Figure 12.
Granulometric distribution of condensed particles depending on the consumption of the initial powder at p = *100 KPa.*

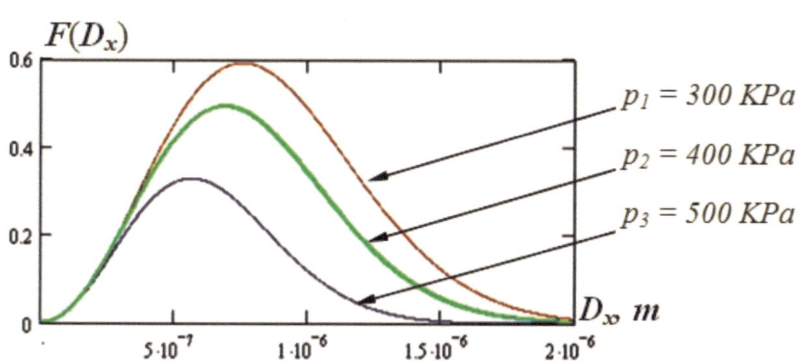

Figure 13.
Granulometric distribution of condensed particles depending on the pressure of the inert gas at G = *0.2 g/min.*

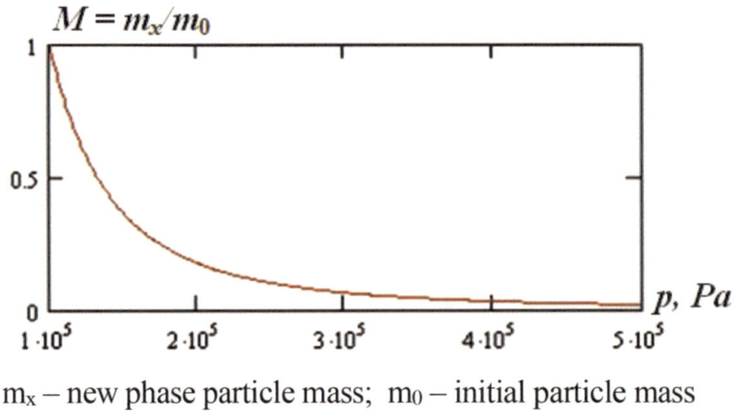

Figure 14.
Relative mass distribution of condensed particles depending on the pressure of the inert gas at G = 0.2 *g/min.*

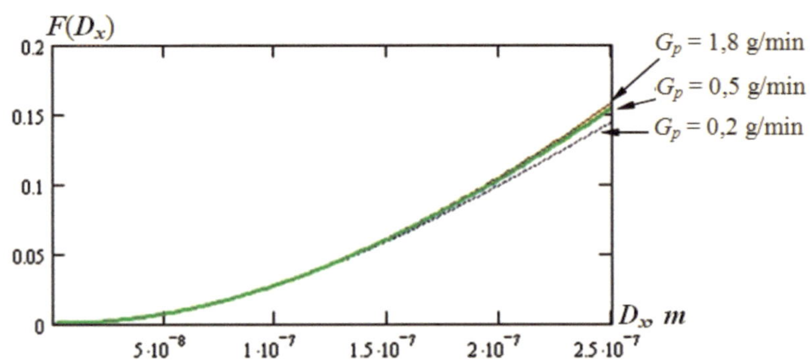

Figure 15.
Granulometric distribution of condensed particles after an ideal filter at 250 nm, depending on the feed rate of the initial powder at p = 100 *kPa.*

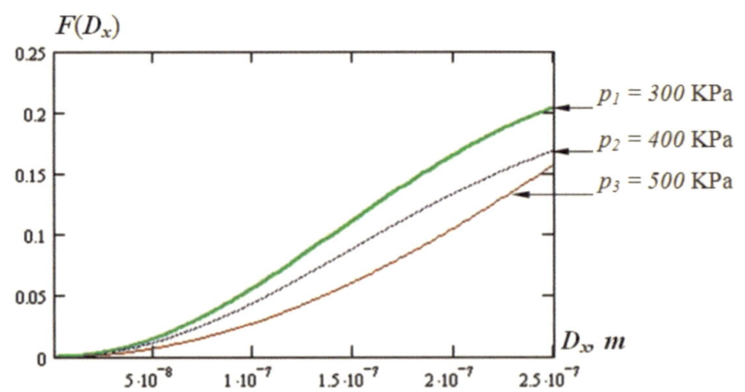

Figure 16.
Granulometric distribution of condensed particles after an ideal filter at 250 nm, depending on the pressure of an inert gas at G = 0.2 *g/min.*

1. On the whole, the granulometric composition and the number of particles of the new phase obtained after ablation are generally affected not only by the energy, pulse duration, and frequency of laser radiation, but also by the pressure of the

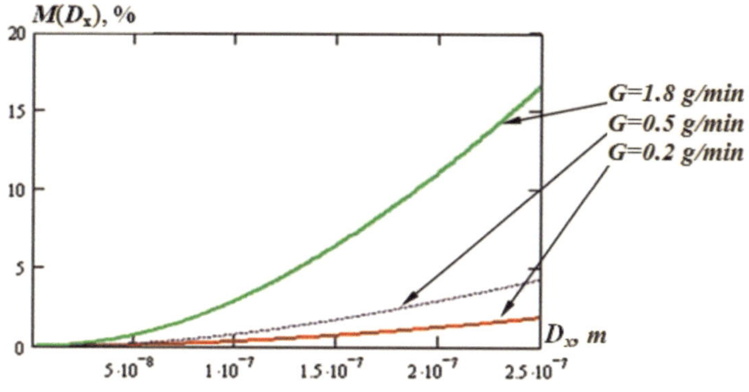

Figure 17.
Distribution of condensed particles by mass after an ideal filter at 250 nm.

inert gas in the zone of action of this radiation, as well as the feed rate of the initial metal powder of the medium fraction (10–100 μm). In this case, an increase in the pressure of an inert gas in the ablation zone has a two to five times less effect on the relative amount of particles of the new phase obtained than the feed rate of the initial powder. If the feed rate of the initial powder is related to the height of the layer that puts pressure on the particles falling out of the dispenser, then the rate of their movement in the laser ablation zone will be the lower, the lower the height of this layer.

2. For spherical titanium particles of fraction D_0 = 45–63 μm, moving at a set speed in the laser ablation zone with a length lpr. = 10 mm, in which, according to the initial conditions, the pulse energy W_{pulse} = 3 J, its duration τ_{pulse} = 3 ms, and the frequency laser radiation following f = 50 Hz, at a feed rate of the initial powder at a level of G = 0.2 g/min after laser ablation, the highest relative (up to 50%) obtaining of particles of a new phase is possible, in which the arithmetic mean diameter has a value of D_x = 800 nm. If under these conditions, due to gas filtration, it is necessary to isolate the largest number of particles of a new phase with a diameter of $D_x \leq 250$ nm, then the feed rate of the initial titanium powder into the laser ablation zone should be set at the level G = 1.8 g/min. Then, the number of particles of the new phase separated after filtration of the required fraction will be nine times greater than at a feed rate of G = 0.2 g/min.

4. Laboratory studies of the possibility of obtaining ultrafine powders of fraction ≤ 250 nm using filtration and laser ablation of vertically falling spherical particles of titanium with a fraction of 45–63 μm

When conducting laboratory studies, a powder laser stereo lithography unit (analogous to a 3D printer) was used as the main technological equipment [13]. The appearance of this analogue of this 3D printer is shown in **Figure 18**, and its characteristics are shown in **Table 3**. As the main equipment in the design of a 3D printer analog, the following were used: a pulsed solid-state laser 1, an optical-mechanical system 2 for focusing and positioning on the plane of the laser beam, a sealed technological chamber 3 with a vertically moving construction platform, a powder dispenser

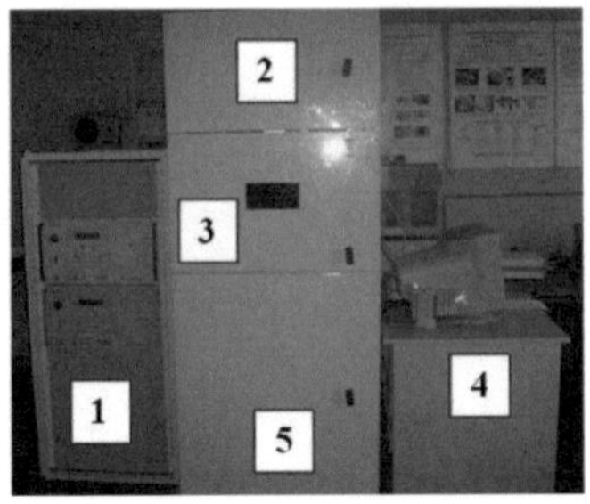

Figure 18.
Appearance of metallurgical 3D printer analogue. 1: Solid-state laser; 2: Optical-mechanical system; 3: Technological chamber; 4: Control computer; 5: Modular compartment.

Characteristics	M. unit	Value
Laser wavelength	nm	1060
Duration of laser radiation	ms	1–3
Laser frequency adjustment range	Hz	1–50
Laser pulse energy adjustment range	J	0,3–3
Operating gas	—	Argon, Nitrogen

Table 3.
Technical characteristics of the metallurgical 3D printer.

and a supply and exhaust system supplying working gas, as well as a control computer 4. To accommodate additional equipment, a modular compartment 5 is provided here.

Due to the block modular design, when conducting research in the technological chamber of a 3D printer analogue according to the scheme shown in **Figure 19**, opposite the lens of the opto-mechanical system focusing on the laser radiation stream 1, a conical laser beam shaper 2 was installed, as well as a powder feeder 3, a dispenser 4, an ultrafine powder sampling device 6, and an FT-3-1109 microfiber filter (Munktell) 7, a hopper for collecting unevaporated powder 8, and a bubbling filter 9.

The studies were carried out using the modes recommended on the basis of the simulation results and contained in the conclusions of sections 1 and 2. According to this, the energy of laser radiation pulses in the ablation zone was W_{puls} = 3 J with their duration τ_{puls} = 3 ms and repetition rate f = 50 Hz. The temperature in the technological chamber filled with argon was T = 50 ^{0}C at a pressure close to normal (120 kPa). The mass flow rate of vertically falling spherical particles of titanium powder into the laser ablation zone was G = 1.8 g/min. The granulometric composition of spherical particles of titanium powder after sieve classification corresponded to a fraction of 45–63 μm. The result of the studies was the isolation (**Figure 20**) of particles of the required fraction of ultrafine titanium powder, the relative mass of which in relation to the initial one did not exceed 14–16%.

The results of laboratory studies of the possibility of isolating by filtration of an ultrafine powder of a fraction ≤250 nm obtained from vertically falling spherical particles of titanium of 45–63 μm fraction using laser ablation in the modes recommended by modeling, concerning the relative amount (14–16%) of the

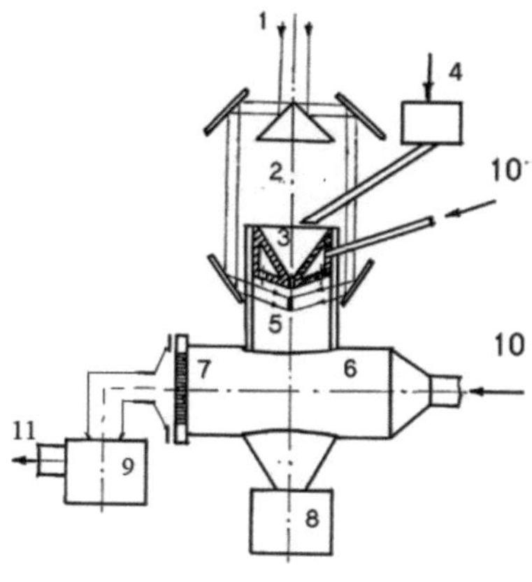

Figure 19.
Scheme of laboratory research. 1: Laser ray; 2: Shaper of conical laser beams; 3: Powder feeding device; 4: Dispenser; 5: Laser ablation zone; 6: Ultrafine powder sampling device; 7: Microfiber filter; 8: Hopper for collecting unevaporated powder; 9: Sparge filter; 10: Argon; 11: To the outlet of the supply and exhaust system supplying working gas.

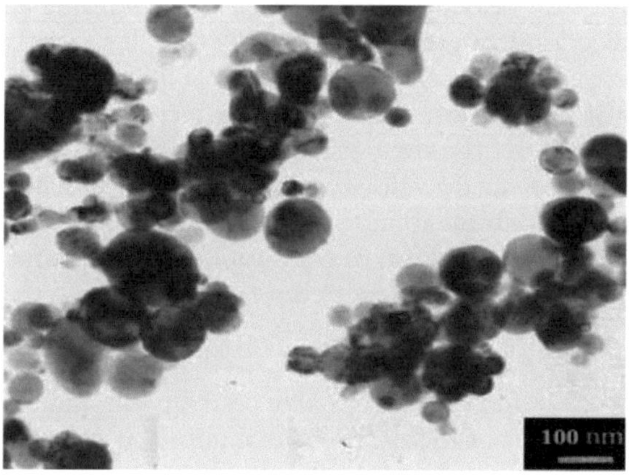

Figure 20.
Electronic diffraction pattern of titanium particles isolated after filtration.

separated fraction with respect to the initial one, are also consistent with the forecast indicators (**Figure 16**) obtained by modeling.

5. Conclusions

1. Using the above research results as an example, we show the fundamental possibility of using the developed foundations of mathematical modeling to determine the necessary modes for obtaining an ultrafine powder by laser ablation of metal particles vertically falling in an inert gas medium under the action of gravity, provided that these particles flow around laminar flows of this gas.

2. Along with this, the following is also shown:

a. II-1. In a free vertical fall under the action of gravity, the particles of the medium and fine fractions of metal powders located in an inert gas experience deceleration due to the emerging buoyancy forces and gas-dynamic resistance, which affects the nature of their movement. A feature of this movement is that the gas flows around the particles are essentially laminar or close to laminar. The speed of the particles, which increases with falling, after a certain period of time reaches the established values.

b. II-2. The path traveled, the time of transition to a constant value of the particle fall velocity, and the value of this velocity itself are influenced not only by the size of the particles, but also by the density, pressure, and temperature of the inert gas in which they move. If argon is an inert gas, then in the selected range of its temperatures T = 0–500°C and pressures p = 100–500 kPa, a change in the first parameter (temperature) will, with a high degree of certainty, lead to an inversely proportional change in the value of the steady velocity, and a change in the second parameter (pressure), on the contrary, to a linear change in this value.

c. II-3. The steady velocity of the particles is also influenced by the pressure of their ejection from the metered feeder. If the particles move in a hollow optically transparent tube into the zone of action of the conical laser beam, and from it into the laser ablation chamber, then it is possible to additionally adjust the time of their transition to the equilibrium state and the value of the steady velocity by varying the pressure, temperature, and flow rate of the supplied inert gas. These parameters include the granulometric composition of the supplied powder and its flow rate, as well as the pressure that pushes out the particles; all these taken together determine the modes of powder supply to the laser ablation zone.

d. II-4. On the whole, the particle size distribution and the amount of titanium particles of the new phase obtained after ablation are generally affected not only by the energy, pulse duration, and frequency of laser radiation, but also by the pressure of the inert gas in the zone of action of this radiation, as well as the feed rate of the initial powder into it. In this case, an increase in the pressure of an inert gas in the ablation zone has a two to five times less effect on the relative amount of particles of the new phase obtained than the feed rate of the initial powder. If the powder feed rate is related to the height of the layer that puts pressure on particles falling out of the dispenser, then the rate of their movement in the laser ablation zone will be the lower, the lower the height of this layer.

e. II-5. For spherical titanium particles of medium fraction D_0 = 45–63 μm, moving at a set speed in the laser ablation zone with a length $l_{pr.=}$10 mm, in which, according to the initial conditions, the pulse energy $W_{imp.=}$3 J, its duration τ_{pulse} = 3 ms, and the repetition frequency of laser radiation is f = 50 Hz, at a feed rate of the initial powder at a level of G = 0.2 g/min after laser ablation, the highest relative (up to 50%) obtaining of particles of a new phase is possible, in which the arithmetic mean diameter has a value of D_x = 800 nm. If under these conditions, due to gas filtration, it is necessary to isolate the largest number of particles of a new phase with a diameter of $D_x \leq 250$ nm, then the feed rate of the initial titanium powder into the laser ablation zone should be set at the level G = 1.8 g/min. Then, the number of particles of the new phase separated after filtration of the required fraction will be nine times greater than at a feed rate of G = 0.2 g/min. These simulation results are consistent with a high degree of confidence with the practical results of laboratory studies.

3. The developed foundations of mathematical modeling can also be used [13] in the design and development of advanced technological processes using laser ablation to obtain advanced powders of fine and ultrafine fractions for additive production, consisting of particles with a spheroidized surface, including the possibility of cleaning the surface of these particles from contaminants.

Author details

A.Ph. Ilyushchanka
O.V. Roman Powder Metallurgy Institute, Minsk, Belarus

*Address all correspondence to: alexil@mail.belpak.by

References

[1] Osokin EN, Artemjeva OA. Powder Metallurgy Processes. Version 1.0 – Electronic Data (5 Mb). Krasnoyarsk: IPK SFU; 2008. p. 421

[2] Zelenko MA, Nagajcev MV, Dovbysh MV. Additive Technologies in Mechanical Engineering. Moscow: NAMI; 2015. p. 220

[3] Dovbysh VM, Zabednov PV, Zelenko MA. Additive Technologies and Metal Products. Moscow: NAMI; 2014. p. 57

[4] Kharazhevsky EV, Krivelev MD. Physics of Lasers, Laser Technologies, Methods of Mathematical Modeling of Laser Action on Matter. Izhevsk: Udmurt University; 2011. p. 187

[5] Arhipov VA, Usanina AS. Movement of Aerosol Particles in a Stream. Tomsk: Publishing House of Tomsk State University; 2013. p. 92

[6] Grozdovsky GL. On the Movement of Small Particles in a Gas Stream. Moscow: TsAGI Science Journal «Uchenye zapiski»; 1974. pp. 80-89

[7] Yavorsky BM, Detlaf AA. Handbook of Physics for Engineers and University Students. Moscow: Nauka; 1979. p. 941

[8] Solonina OP, Glazunov SG. Heat-resistant titanium alloys. Moscow: Metallurgy; 1976. p. 448

[9] Ilyin AA, Kolachev BA, Polkin IS. Titanium Alloys. Composition, Structure, Properties. Moscow: VILS-MATI; 2009. p. 520

[10] Kostornov AG. Material Science of Dispersed and Porous Metals and Alloys. Vol. 1. Kyiv: Naukova Dumka; 2002. p. 572

[11] Syzrantsev VN, Nevelev YP, Golofast SL. Adaptive methods for restoring the probability density function. News of higher educational institutions. Engineering. 2006;**12**:3-11

[12] Voronin GF. Fundamentals of Thermodynamics. Moskow: MSU; 1987. p. 192

[13] Ph A. Ilyushchanka, Additive Technologies and Powder Metallurgy. Minsk: Medisont; 2019. p. 260